Advances in Industrial Control

For further volumes:
http://www.springer.com/series/1412

Ali Zolghadri • David Henry • Jérôme Cieslak
Denis Efimov • Philippe Goupil

Fault Diagnosis and Fault-Tolerant Control and Guidance for Aerospace Vehicles

From Theory to Application

Springer

Ali Zolghadri
CNRS, IMS Lab
The University of Bordeaux
Bordeaux
France

David Henry
CNRS, IMS Lab
The University of Bordeaux
Bordeaux
France

Jérôme Cieslak
CNRS, IMS Lab
The University of Bordeaux
Bordeaux
France

Denis Efimov
INRIA-LNE
Non-A project, Parc Scientifique
de la Haute Borne
Villeneuve d'Ascq
France

Philippe Goupil
EYCCC Flight Control System
Airbus Operations S.A.S. – St. Martin
Toulouse
France

ISSN 1430-9491 ISSN 2193-1577 (electronic)
ISBN 978-1-4471-5312-2 ISBN 978-1-4471-5313-9 (eBook)
DOI 10.1007/978-1-4471-5313-9
Springer London Heidelberg New York Dordrecht

Library of Congress Control Number: 2013945429

Printed on acid-free paper

Springer is part of Springer Science+Business Media (www.springer.com)

Series Editors' Foreword

The series *Advances in Industrial Control* aims to report and encourage technology transfer in control engineering. The rapid development of control technology has an impact on all areas of the control discipline. New theory, new controllers, actuators, sensors, new industrial processes, computer methods, new applications, new philosophies . . . , new challenges. Much of this development work resides in industrial reports, feasibility study papers, and the reports of advanced collaborative projects. The series offers an opportunity for researchers to present an extended exposition of such new work in all aspects of industrial control for wider and rapid dissemination.

Recently the sister series to *Advances in Industrial Control*, the *Advanced Textbooks in Control and Signal Processing* series published a textbook, *Robust and Adaptive Control: With Aerospace Applications* (ISBN 978-1-4471-4395-6, 2012). This was written by two Boeing Senior Technical Fellows, Doctors Eugene Lavretsky and Kevin A. Wise. This textbook was notable for a number of reasons: firstly, it was written by aeronautical engineers from within the industry. Secondly, the textbook demonstrated the use of advanced robust and adaptive control techniques for aerospace applications; this shows how successfully advanced control techniques are beginning to penetrate an industry with very strict certification procedures.

This *Advances in Industrial Control* monograph, *Fault Diagnosis and Fault-Tolerant Control and Guidance for Aerospace Vehicles: From Theory to Application* by Ali Zolghadri, David Henry, Jérôme Cieslak, Denis Efimov, and Philippe Goupil, has similar strengths in that it reports work with advanced fault detection and control techniques but is strongly grounded in commercial aeronautical practice. Throughout the monograph are important sections detailing actual current aeronautical in-service practice, and various demonstrations are presented using, for example, Airbus benchmark models and simulations. This industrial involvement arises from the very nature of European collaborative research projects that must have the critical participation of industrial companies. Many academic and industrial readers will find this monograph a refreshing change from the more usual academic control monographs that are often so dependent on idealized process models in the examples.

Opening the monograph is an introductory chapter (Chap. 1) followed by Chap. 2 that is a review chapter. The review material is comprehensive encompassing the industrial state of the art and the academic state of the art in fault detection and related control topics.

The first three core Chaps. (3, 4, and 5) focus on fault detection and diagnosis (FDD) systems that answer the questions of whether there is a fault present and then what exactly that fault is. Model-based theories for FDD and Kalman estimation are two of the methods invoked in the proposed solutions in these chapters. This is followed logically by the development of recovery control strategies, in this case based on fault-tolerant control (FTC) methods where H_∞ design is a feature of the proposed control solutions (Chap. 6).

Chapter 7 moves the application focus into outer space with three application scenarios being investigated: an observation satellite in Earth's orbit, an atmosphere reentry vehicle, and a deep space mission. All the applications are linked to European space research projects; so once again there is a valuable association with the reality of ongoing space projects and research. The proposed solutions feature H_∞ design tools. Closing the monograph, Chap. 8 presents conclusions and a perspective on future research directions.

This monograph showcases much research arising from a number of European and French research projects in the aeronautics/aerospace field. The strong underpinning link with industrial activities in this area makes for a strongly applications-oriented text that sits very well within the remit of the *Advances in Industrial Control* monograph series.

Industrial Control Centre
Glasgow
Scotland, UK

M.J. Grimble
M.A. Johnson

Foreword

An important aspect related to airplane flight control is the movement of surfaces on the wing and on the rear tailplane and fin. Wrong movements of these control surfaces may have consequence on the trajectory of the airplane, on the loads of its structure, or on its fuel consumption. Detecting and passivating these wrong movements is absolutely needed. However, the failure detection devices must not cry wolf untimely. False alert to the crew can disturb him from more essential activities. Wrong failure detection may lead to disconnect (automatically or not) resources that could be needed later on in the flight and in any case need to be fixed later on after landing, and this is a very heavy burden for the airplane operators. Monitoring devices must thus be tolerant to the normal behavior of the systems (in particular the normal tolerances of sensors, propagated by the functions that are consuming their output) but must detect any signal that could impair airplane safety.

Failure detection mechanisms must be squeezed between normal behavior and safety constraints. Safety constraints can be alleviated by introducing margins in the trajectory of the airplane (increasing separation in flight between airplanes), margins in the sizing of the structure, and margins in the amount of fuel to be loaded in the tanks. These margins are costly in terms of fuel burn, thus both for our environment and for the airlines' bottom line.

Any progress in the art of developing fault detection mechanisms is thus welcomed and is even a must for the aviation industry. One should be careful that this is not an easy task. This issue of trustworthy fault detection is not new: it is valid for traditional mechanically controlled airplane and the numerous actuators and electronics they encompass. However, the issue has significantly grown with the introduction of flight-by-wire airplane and the variety of errors a signal can support: wider frequency range of oscillations, dynamic of data runaway, or hardover. Moreover, having design methods is just a part of the story. Building confidence in these monitoring devices is also mandatory. The art of developing them must thus plan their validation from the start and must rely on well-proven design methods. The art of fault detection needs thus to provide trustworthy development methods that we can justifiably trust and to continuously evolve and improve.

This book is particularly valuable, bringing together both imaginative theories and practical applications. Philippe Goupil and his academic colleagues have done here a continuous and fruitful collaborative effort to bring the best expertise on this complex issue.

Airbus Operations S.A.S.
Toulouse, France

Dr. Pascal Traverse

Preface

The impact that the aerospace and aviation industry has on today's modern society and world's economy is very prominent. As such, the aerospace industry continues at the forefront of engineering research and development technologies. The sector needs continuous improvement including insertion of new technologies. Generally, new technologies are adopted only when there is a clear need in terms of cost or performance benefit: any modification to the existing in-service and already proven technical solutions should be motivated, first of all, by a real industrial need. The technology driver is mainly the market pull. On the other hand, despite potential cost and performance advantages, new methods and technologies entail risk and thus must undergo extensive development, validation, and verification before they can be transitioned to real-world systems. This is especially true for aerospace and aircraft systems. Recent developments in control engineering have had attractive potential for resolving numerous issues related to guidance, navigation, and control (GNC) of flight vehicles. Satisfying the more and more stringent flight requirements requires innovative Fault Detection, Identification, and Recovery (FDIR) approaches and mechanization schemes which can help achieve improved flight performance and reliability, self-protection, and autonomy. The challenges range from predesign and design stages for upcoming and new programs to the improvement of the performance for in-service flying systems. Many future space missions will require increased onboard autonomy including fault diagnosis and the subsequent control and guidance recovery actions. Autonomy supports cost-effective accomplishment of mission goals, and space missions lacking onboard autonomy will be unable to achieve the full range of advanced mission objectives. On the other hand, one of the main issues for the development of future aircraft programs is to improve "green transport," that is, to provide society with an air transport that leaves a smaller carbon footprint. Sustainable air transport will be a serious worldwide challenge, given the anticipated increase in traffic volume and continuing expansion of the world's aviation network with greater aviation connectivity. This will need continuing technological progress in the design of all aircraft systems: airframes, propulsion systems, airborne systems, software and hardware, communications, navigation, control and guidance, etc. At first sight, the link between innovating

FDIR technologies and sustainable development of air transport may not seem obvious. Yet, early and robust diagnosis of faults that have an influence on structural loads could contribute to the overall optimization of aircraft design and so to weight saving for better overall performance in terms of fuel burn, noise, range, and environmental footprint. Putting innovating aircraft FDIR in this perspective can be an important driving factor for its future developments. This will help anticipate the more and more stringent requirements which will come in force for future and more environmentally friendlier programs.

This book focuses on design and analysis of advanced and viable FDIR technologies for aerospace vehicles. The term "viable" covers here some important aspects which are often underestimated in the classical academic literature: tuning, complexity of the design, real-time capability, modularity and possibility to "reuse" or "build around it," evaluation of worst-case performance, robustness in harsh environment, etc. Unfortunately, the lack of consideration of the above issues has led to a widening gap between the advanced scientific methods being developed by the academic control community and technological solutions demanded by the aerospace industry. While the research in all aspects of model-based FDIR went forward since early 1970s, the design methodology involving feasibility analysis and real-world requirements specification is still missing. This is a major reason for the slow progress in applying advanced model-based FDIR at the GNC level of flight vehicles.

The developments offered in this book are based on the authors' experience and lessons learned through their involvement in a number of aerospace research projects with major academic and industrial actors in Europe over the past few years. The chapters are mostly organized according to a "sandwich" model: concrete-theory-concrete. That is, we will motivate the chapter with a specific aerospace application, work out the theory, and finally return to the specific concrete problem. I believe that this model is most useful as it provides clear operational procedures under the conditions that are explicitly stated.

My first thanks go to my coauthors. Within a very inspiring teamwork, their valuable and everyday work and effort contributed very much to the fascinating topics covered in this book. The last author (Dr. Philippe Goupil) is with Airbus Operations S.A.S., Toulouse, France, where he is in charge of FDIR activities. I am very grateful to him for giving us continuous precious support and for his patient explanations about flight FDIR technologies, the today industrial constraints, and the future needs. Over the past few years, he played a major role in Europe to bridge the gap between industrial aircraft world and the academic control community in order to pave the way for successful and innovating solutions for future aircraft systems.

I would also like to thank all researchers who contributed, in one way or another, when they were in my research team in Bordeaux. Among others, Dr. Denis Berdjag during his postdoc position, Dr. Efrain Alcorta Garcia during his sabbatical year in Bordeaux, and my (ex) PhD students Anca Gheorghe and Alexandre Falcoz.

I also wish to thank Oliver Jackson and Charlotte Cross at Springer for their precious assistance, Professor Michael Johnson for his useful comments, and Professor Mike Grimble.

Finally, I would like to extend my thanks to Brigitte, Tania, and Sacha Zolghadri for their helpful tips and suggestions.

Bordeaux, France
March 2013

Ali Zolghadri

Contents

Chapter 1
Introduction

1.1 Motivations

This book presents a number of advanced fault detection and diagnosis and reconfiguration technologies for aerospace vehicles. An attempt is made to develop useful solutions that can be relevant and viable candidates for future space and aeronautical systems. The presented techniques have been tested and validated on highly representative benchmarks, real flight data, or real-world aerospace systems. The examples presented in this book are taken mainly from four recent projects related to fault detection and diagnosis and fault-tolerant control and guidance of aircraft and space systems:

GARTEUR Project

GROUP FOR AERONAUTICAL RESEARCH AND TECHNOLOGY IN EUROPE
FRANCE · GERMANY · ITALY · THE NETHERLANDS · SPAIN · SWEDEN · UNITED KINGDOM

From 2004 to 2008, a research group on fault-tolerant control, comprising a collaboration of 13 European partners from industry, universities, and research institutions, was established within this cooperative program. The aim of the research group, Flight Mechanics Action Group (FM-AG) (16), was to demonstrate the viability and performance of innovative reconfigurable flight control algorithms to improve aircraft survivability during upset flight conditions. The group facilitated the proliferation of new developments in fault-tolerant control design within the European aerospace research community towards practical and real-time operational applications. This addresses the need to aid the crew to recover from adverse conditions induced by (multiple) system failures and damage that would otherwise be potentially catastrophic on mechanical flight control system aircraft.
See http://www.nlr.nl/documents/GARTEUR_AG16_Workshop/.

A. Zolghadri et al., *Fault Diagnosis and Fault-Tolerant Control and Guidance for Aerospace Vehicles: From Theory to Application*, Advances in Industrial Control,
DOI 10.1007/978-1-4471-5313-9_1,

ADDSAFE Project: ***Advanced Fault Diagnosis for Sustainable Flight Guidance and Control***

ADDSAFE was a European collaborative project supported by the European Seventh Framework Program (2009–2012). The overall aim was to research and develop model-based fault detection and diagnosis methods for aircraft flight control systems faults, predominantly sensor and actuator malfunctions. The results were intended to help achieve the European Vision 2020 challenges related to the "greening" of the aircraft, by supporting the application of already developed sustainable solutions and by opening the door to develop new technologies while keeping the current highest aircraft safety levels regardless of the increase in air traffic.
See http://addsafe.deimos-space.com.

SIRASAS Project: ***Innovative and Robust Strategies for Spacecraft Autonomy***

SIRASAS was a French collaborative project on spacecraft autonomy (2007–2010). The project gathered together industrial and academic partners to promote innovative and robust technologies that could significantly increase spacecraft autonomy. This project addressed the model-based fault detection, identification, and recovery challenges for G&C (Guidance and Control). The actions undertaken within SIRASAS aimed at overcoming the dead zone between the scientific advanced methods advocated by the academic and research communities and the technological solutions demanded by the aerospace industry, with stringent operational constraints.
See https://extranet.ims-bordeaux.fr/External/SIRASAS/accueil.php.

SICVER Project: ***Fault Detection and Diagnosis and Fault-Tolerant Guidance for Atmospheric Reentry Vehicles***

SICVER was a collaborative research project which has been supported and funded by the European Space Agency and EADS Astrium Space Transportation (2006–2009). The project aimed at developing innovative fault detection and diagnosis and fault-tolerant guidance strategies for experimental reentry vehicles. The project included two research areas. The first one dealt with the design of onboard fault-tolerant guidance ensuring a high level of spacecraft autonomy. The goal was to help the ground-level operations and to improve decision making. The second part dealt with the design of actuator onboard fault detection and diagnosis during an atmospheric reentry mission.

1.2 Book Outline

This book is organized in eight chapters.

Chapter 2: This chapter starts with some basic definitions and concepts as well as a quick literature review on FDIR academic methods. The main concepts of the industrial state-of-practice for space and avionics systems will also be briefly presented. An attempt will be made to analyze major reasons for the slow progress in applying advanced model-based techniques to real-world aerospace systems.

Chapter 3: This chapter deals with model-based fault detection and diagnosis (FDD) methods which have been recently applied to Oscillatory Failure Case (OFC) in aircraft control surface servo-loops. This failure case, related to the electrical flight control system (EFCS), could have an influence on structural loads and aircraft controllability. Two methods will be presented and, in order to improve FDD performance and robustness, the tuning of their free design parameters are discussed. The presented methods are nonlinear observer design and fault reconstruction via sliding-mode differentiation. The efficiency of the above techniques will be illustrated through their application to highly representative aircraft benchmarks, real flight data, and real-time implementation on Airbus test facilities.

Chapter 4: This chapter is dedicated to two other important EFCS-failure cases in aviation: runaway and jamming. A runaway is an untended (or uncontrolled) deflection of a control surface which can go until its stops if it remains undetected. A jamming is a scenario where a control surface is physically stuck at its current position. It will be shown that by careful fault modeling, simple estimation techniques (Kalman-based) can lead to remarkable results. The technique has been implemented as a part of the A380 flight control computer (FCC) software and provided very good results on the Airbus test facilities. The robustness of the method has been confirmed during about 70 h of flight tests.

Chapter 5: This chapter is dedicated to techniques for ensuring fault tolerance in redundant aircraft sensors involved in computation of flight control laws. The objective is to switch off the faulty sensor and to compute a reliable (aka as "consolidated") parameter using data from valid sensors, in order to eliminate any anomaly before propagation in the control loop. The benefit of the presented method is to improve the consolidation process with a fault detection and isolation approach when only few sources (less than three) are valid. Different techniques are compared to accurately detect any behavioral change of the sensor outputs. The approach is tested on a recorded flight data set.

Chapter 6: This chapter deals with the next step following the design of an FDD system, i.e., appropriate recovery strategies, based on all available actuator/sensor/communication resources. An active fault-tolerant flight control strategy based on H_∞ design tools is presented. The fault-tolerant control (FTC) strategy operates in such a way that once a fault is detected and confirmed by a FDD unit, a compensation loop is activated for safe recovery. A key feature of the proposed strategy is that the added FTC loop keeps unchanged the in-service control laws facilitating the certification of the whole approach and limiting the underlying verification and validation activities. The methodology is applied to actuator fault accommodation of a large commercial aircraft during landing approach. The results, obtained from a piloted 6-DoF flight simulator, will be presented and discussed. The application is taken from the GARTEUR project.

Chapter 7: This chapter is dedicated to space applications. Three application cases will be presented: an Earth observation satellite, a deep space mission, and an atmospheric reentry vehicle. The design method is based on H_∞/H_- tools and is associated with a suitable post-analysis process, the so-called generalized μ-analysis. It is shown that the resulting design/analysis procedure provides an iterative refinement cycle which allows the designer to get "as close as possible" to the required robustness/performance specifications and trade-offs.

Chapter 8: This chapter is dedicated to final remarks and suggestions on future challenges and opportunities. We focus on what useful and realistic contributions can the research community make in order to develop successful solutions for future space and avionics systems.

Chapter 2
Review and Basic Concepts

Acronyms

EFCS	Electrical Flight Control System
FBW	Fly-by-Wire
FCC	Flight Control Computer
FDD	Fault Detection and Diagnosis
FDI	Fault Detection and Isolation
FDIR	Fault Detection, Identification and Recovery
FTC	Fault-Tolerant Control
FTG	Fault-Tolerant Guidance
L/D	Lift-to-Drag Ratio
NEP	Nominal Exit Point
GNC	Guidance, Navigation, and Control
HMI	Human–Machine Interface
LTI	Linear Time Invariant
LPV	Linear Parameter Varying
RLV	Reusable Launch Vehicle
TAEM	Terminal Area Energy Management
TEP	TAEM Entry Point
TRL	Technology Readiness Level
α	Angle-of-Attack
M	Mach Number

A. Zolghadri et al., *Fault Diagnosis and Fault-Tolerant Control and Guidance for Aerospace Vehicles: From Theory to Application*, Advances in Industrial Control,
DOI 10.1007/978-1-4471-5313-9_2, © Springer-Verlag London 2014

2.1 Introduction

2.1.1 Fault Detection and Diagnosis, Fault-Tolerant Control, and Fault-Tolerant Guidance

Fault detection and diagnosis (FDD) is an important aspect of process engineering. The primary objective of an FDD system is early detection of faults, isolation of their location, and diagnosis of their causes, enabling correction of the faults before additional damage to the system or loss of service occurs. Abnormal situations occur when processes deviate significantly (outside the allowed range) from their normal regime during online operation. A fault can be defined as an unpermitted deviation of at least one characteristic property or parameter of the system from the standard condition [1]. A failure is a permanent interruption of a system's ability to perform a required function under specified operating conditions. Within the academic literature, the terminology is now more or less standardized.[1] Such malfunctions may occur in the individual unit of the plants, sensors, actuators, or other devices and affect adversely the local or global behavior of the system. Process abnormalities are usually classified into additive or multiplicative faults according to the effects on a process. In general, additive faults affect processes as unknown inputs, while multiplicative faults usually have important effects on the process dynamics and can cause unstable behaviors. Abrupt faults are sudden changes in behavior of the system (step like), while incipient faults are gradual and slow drifting faults. Permanent faults lead to the total failure of the equipment (once they occur they do not disappear), transient faults are temporary malfunctioning (appear for a short time and then disappear), and intermittent faults are the repeated occurrences of transient faults (they appear, disappear, and then reappear). Hidden faults are those which are present on standby equipment and visible only when this equipment is activated.

Throughout this book, we do not consider software and communication bugs for which the detection techniques are very different from the techniques used to handle physical faults.

Generally, the main desirable characteristics of an FDD system are:

- Early detection and diagnosis, i.e., detection delay should be minimized. This feature is highly related to the fault/failure criticality.
- Good ability to discriminate between different failures (isolability).
- Good robustness to various noise and uncertainty sources and their propagations through the system.
- High sensitivity and performance, i.e., high detection rate and low false alarm rate.

---

[1]See, for example, http://www.safeprocess.es.aau.dk/

Once faults are correctly detected, confirmed, and diagnosed, a reconfiguration mechanism may be used in order to achieve fault tolerance. The primary goal of fault tolerance is to prevent errors from propagating and leading to a dangerous, hazardous or off-normal system behavior. For many safety-critical systems, fault tolerance is founded on redundancy. If we have two or more identical components, we can ignore the faulty component or switch to a spare if the primary fails. For flight systems, recovery and reconfiguration actions may have different goals and characteristics depending on the considered mission. For example, for an observation satellite, reconfiguration mechanisms are based on redundant units switching and consist in action sequences, i.e., event sequences or onboard control procedures, which are a priori programmed and then executed as a reflex reaction following fault detection [2].

From a "control" point of view, one can distinguish two basic functions for reconfiguration: fault-tolerant control (FTC) and fault-tolerant guidance (FTG). FTC systems seek to provide, at worst, a degraded level of performance in the faulty situations. Generally, a fault-tolerant control does not offer optimal performance for normal system operation, but it can compensate effects of system failures by adjusting, for example, the controller parameters to recover the system from the faulty condition. In general, FTC strategies are classified into passive and active approaches. In the passive approach, a single control law is designed to keep stability and an acceptable level of performance in both fault-free situation, i.e., when all components are operational, and in the case of faults. It can be seen as a "super" or augmented robust control law. The price to pay for robustness to faults is that nominal and fault-free performance is deteriorated. An active FTC strategy requires FDD information for control reconfiguration (see, e.g., [3–5]). FTG could provide a greater flexibility for safe recovery in case of degraded flight conditions. In fact, onboard planning capabilities can be used to resume mission activities without ground intervention after a fault is detected and confirmed. It supposes a diagnosis capability and the possibility to take into account deteriorated resources in the planning process. FTC and FTG provide means to avoid and suppress a potentially hazardous, out-of-tolerance, or dangerous behavior of the system if possible or provide means by which the consequences of a dangerous behavior are avoided.

2.1.2 Interaction Between FDD, FTC, and FTG

Conceptually, the interaction between FDD, FTC, and FTG units can be illustrated as in Fig. 2.1. FTC follows FDD and provides means to continue to "control" the faulty system (maintain stability and achievable performance). FTG would be necessary when the available onboard control resources are limited and when FTC would not be sufficient.

For aerospace applications, the above functions are related to the GNC (guidance, navigation, and control) system. The GNC system gives the vehicle the ability to execute flight over a predefined path generated by a path planner. Guidance

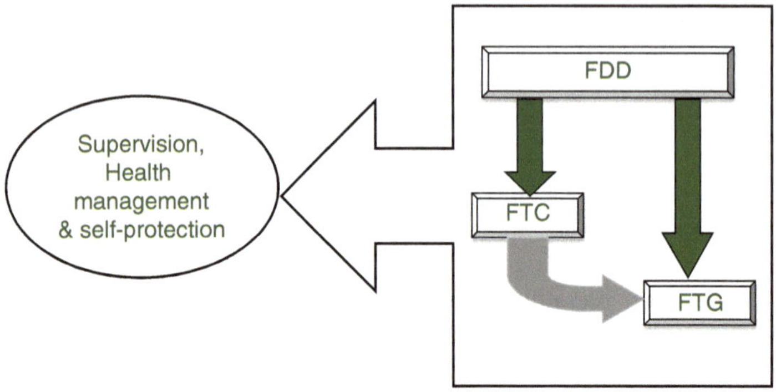

Fig. 2.1 Interaction between FDD, FTC, and FTG

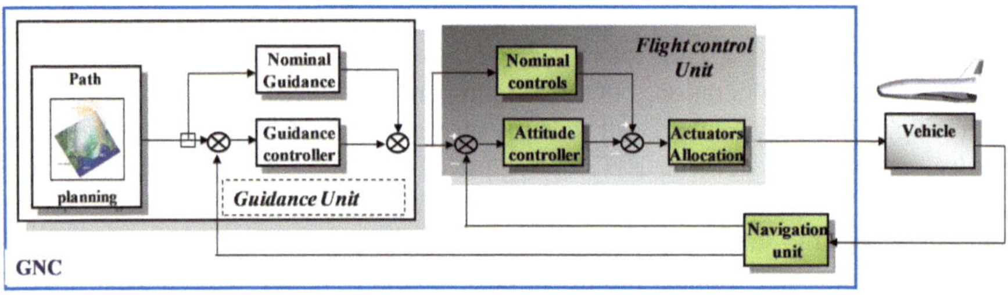

Fig. 2.2 GNC system

equipment (gyroscopes, accelerometers . . .) compute the location (or attitude) of the vehicle and the orientation required to satisfy mission requirements. Navigation tracks the vehicle's actual location and orientation. Usually control consists of two modes: automatic and manual. In the automatic mode, the primary avionics software system allows the onboard computers to control the guidance and navigation of the space vehicle. In the manual mode, the flight crew uses data from the GNC displays and hand controls for the guidance and navigation. Although GNC design is by far the most relevant aspect for aircraft and space vehicles, its treatment is well beyond the aim of this book. The interested reader can refer to many published materials on this subject, among others the dedicated conferences organized by AIAA (https://www.aiaa.org).

A simplified block diagram of the GNC is depicted in Fig. 2.2. Using air data and engine thrust data, the guidance loop computes the guidance demands to follow waypoint scenarios. The flight control loop generates actuator signals for the control surfaces. As aerospace vehicles are often over actuated, a control allocation (often static, sometimes dynamic) allows for distributing a desired total control effort among a redundant set of actuators.

A more detailed description of the FDD and FTC functions will be given in the following sections. Roughly speaking, FTG means "change the mission objectives." To illustrate the idea of FTG, consider a typical atmospheric reentry trajectory

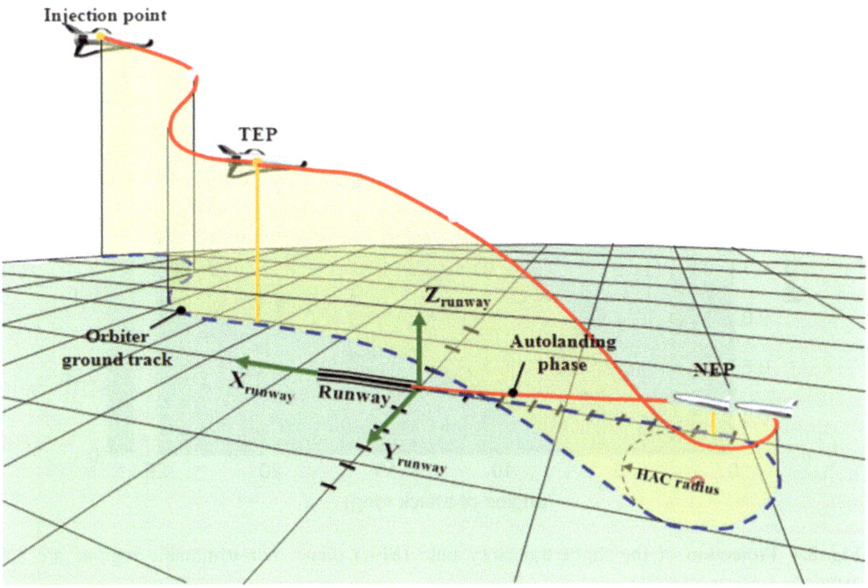

Fig. 2.3 Atmospheric reentry trajectory

(Fig. 2.3) for a medium- or high-L/D vehicle. It consists in performing three successive flight phases, namely, the hypersonic phase from about 120 km high down to TAEM handover, the TAEM phase from Mach 2 gate down to Mach 0.5 gate, and the auto-landing phase from Mach 0.5 gate down to the wheel stop on the runway. After having achieved the hypersonic path, the vehicle initiates the TAEM phase characterized by an entry point, called TEP, typically defined when crossing the Mach 2 gate and an exit point, called NEP, which is defined in terms of altitude, velocity, and distance to the runway. Finally, the landing path is defined in terms of desired altitude from the runaway threshold, and it is composed of three successive sections, i.e., a steep outer glide slope, parabolic pull-up maneuvers, and a shallow inner glide slope. During the reentry mission, actuator failures and damage of control effectors could lead to substantial performance degradation and even instability of the closed-loop system. An important issue following the FDD consists then to engage timely safe recovery actions to accommodate faults. The goal is to maintain control of the vehicle following actuator faults by means of the healthy control effectors. However, under some failure conditions, even advanced FTC techniques may be insufficient to recover the vehicle. Significant aerodynamics characteristic change of the vehicle and a possible lack of control may require reshaping of a new trajectory so as to land the vehicle safely and in compliance with the stringent operational and fight dynamics constraints. Key features for the success of such reshaping algorithms rely on the knowledge of the failed actuator position (reliable FDD information) so as to evaluate the remaining capabilities of the vehicle to be rotationally trimmed. The case of non-compensable faults can

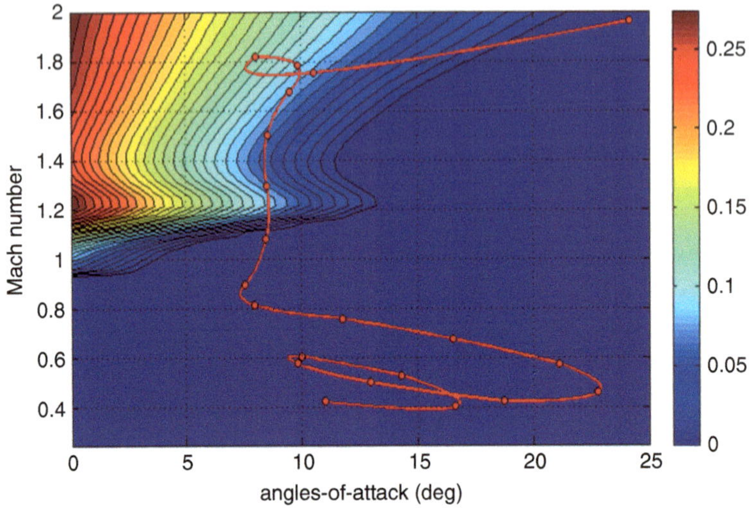

Fig. 2.4 Projection of the flight trajectory onto (M-α) plane: non-trimmable regions are not avoided

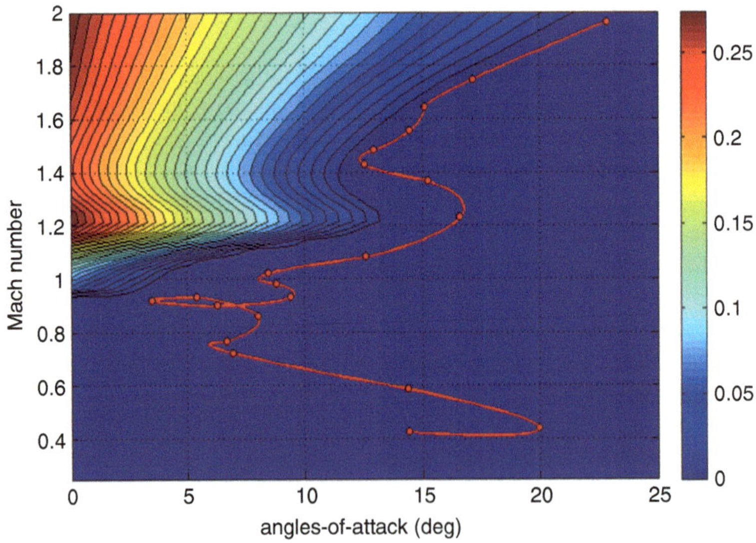

Fig. 2.5 Successful FTG: reshaped trajectory

be then studied as a trimmability-deficiency analysis problem which boils down to a static-fault compensability study. The goal is to define the flight envelope regions, for example, in the Mach-α space (see Figs. 2.4 and 2.5), where the vehicle cannot be rotationally balanced in the presence of faults. As a direct consequence, a fault is considered as non-compensable if the flight trajectory of the vehicle (projected in the Mach-α space) crosses the non-trimmable region (see Fig. 2.4 for an illustration). It follows that the results after a successful FTG corresponds to a flight trajectory that does not cross the non-trimmable region (see Fig. 2.5 for an illustration). In Figs. 2.4 and 2.5, both flight trajectory (red) and trimmability-deficiency regions (from blue = trimmable regions to red = highly non-trimmable regions) are depicted.

2.1.3 Chapter Organization

This chapter is organized as follows. Section 2.2 presents a brief overview of the industrial state-of-practice. Section 2.3 is devoted to the review of the available academic literature. Section 2.4 highlights the reasons for slow-developing progress of the advanced academic methods to real-world aerospace systems. Finally, Sect. 2.5 is dedicated to final remarks and motivates the developments that will be presented in subsequent chapters.

2.2 Industrial State-of-Practice

2.2.1 General Ideas

The basic principles involving general health management architecture trade-offs changed little from the 1960s, although the hardware mechanizations of the earlier analog systems have been replaced largely with the software of the newer digital systems (see, e.g., [6, 7] for a historical review). The success of the Apollo program has been an important factor for the development of digital fly-by-wire technologies. In the late 1960s, engineers at NASA Flight Research Center (now NASA Dryden) proposed replacing bulky mechanical flight control systems on aircraft with much lighter weight and more reliable analog fly-by-wire technology. As the Apollo program came to completion in the early 1970s, NASA Dryden engineers developed a digital fly-by-wire solution using the specialized software and hardware developed for Apollo [7, 8]. A few years before in Europe, Aerospatiale (now EADS) engineers developed and installed the first analog electrical flight control system on Concorde.[2] In civilian and military aviation, this precipitated a revolution in aircraft design. The electrical flight control system, designed with digital technology on Airbus aircraft from the 1980s (on A310 aircraft for the spoilers, slats, and flaps only and then generalized on all control surfaces on the A320 in 1987), provided more sophisticated control of the aircraft and flight

[2] A supersonic passenger airplane jointly developed and produced by Aerospatiale (France) and the British Aircraft Corporation under an Anglo-French treaty (first commercial fly in 1969).

envelope protection functions. Physical separation of critical avionics functions from less critical functions has been always the primary strategy used by the designers of civil aircraft to produce safe avionic systems. Traditional avionics systems are built around federated architectures in which each processing site contains a single application such as an autopilot, flight management system, or display. Critical functions are protected from noncritical tasks by physical isolation. NASA used this approach on interplanetary spacecraft, where critical functions to the survival of the spacecraft are handled by an attitude and articulation control system, which is separate from the systems that control the science experiments.

Fault detection is generally based on the concept of redundancy, i.e., the comparison of duplicative signals generated by various hardwares, such as measurements of the same parameter given by two or more identical sensors. Fault detection and confirmation is mainly performed by cross-checks, consistency checks, voting mechanisms, and built-in test techniques (BIT, which include hardware sensors and software error correcting codes) of varying sophistication. The typical method for this is limit checking, i.e., verifying whether a parameter value goes outside of a specified range of values. Multiple ranges can be defined; one can specify a not-to-exceed value (high limit) and a low limit. Multiple high or low limits can be also specified, for example, an advisory range, a caution range, and a warning range. The limit can be applied directly to the instantaneous value of the parameter, the change from the previous value, or the trend of the value over time. For instance, a typical commercial aircraft's navigation sensing system can contain triple-redundant inertial references plus triple-redundant air data sensors. A voting scheme monitors and checks the performance of the individual sensors and detects abnormal behavior. A key issue relates to definition of failure thresholds, which reflect calibration tolerances and environmental effects on component specifications. Flight condition-based thresholds, once validated with all the known delays and uncertainties in the signal propagation (acquisition, processing ...), are used for rapid recognition of out-of-tolerance conditions. The main advantage of fixed thresholds is that it allows designers and operators to use and manage them easily. In setting these thresholds, compromises have to be made between the detection size of abnormal deviations and false alarms because of normal fluctuations of the variables.

Fault tolerance relies mainly on hardware redundancy, safety analysis, dissimilarity, physical installation segregation, and hardware/software reconfiguration [9]. For a general analysis of fault-tolerance management in space vehicles, see, for example, [2, 7]. These hardware-based redundancy techniques are nowadays the standard industrial practice and fit also into current industrial certification processes while ensuring the highest level of safety standards.

2.2.2 *Aeronautics*

Firstly, let us look at fault management procedures in cockpit and flight deck. The today flight deck represents a highly automated mass of complex systems with which the flight crew has to interact. The increase in automation has shifted the role of the pilot away from hands on flying and more toward system monitoring. Pilots rely on warning systems to generate alert messages at the earliest opportunity in order to allow maximum time for corrective actions. Each warning has an associated procedure. These procedures are listed in the *Quick Reference Handbook* and *Flight Operations Manual* on the flight deck or, in some cases, are displayed electronically. Basically, all alert messages can be plotted onto two axes: intervention immediacy and intervention importance. These two factors combined establish the alert's urgency. The situation being monitored is often complex with many components, influences, and interactions, and there is a need to take into account a large number of parameters in order to assess the situation see [10–13].

The paper [9] focuses on a typical Airbus EFCS and provides a detailed description on the industrial practices and strategies for FTC and FDD in civil aircraft. Today, the EFCS constitutes an industrial standard for commercial aircraft applications. It provides sophisticated control of the aircraft and flight envelope protection functions [14, 15]. The main characteristics are that high-level control laws in normal operation allow all control surfaces to be controlled electrically and that the system is designed to be available under all possible external disturbances. The EFCS is designed to meet very stringent requirements in terms of safety and availability, specified by the aviation authorities [16]. Compared to mechanical flight control system, it has brought more safety, increased performance, more availability, weight saving, a more accurate control, and an easiest way to update the whole system. However, the EFCS development on modern civil aircraft also led to a growing complexity of systems and equipment. Consequently, the number of failure cases to consider in the aircraft design has increased compared to the historical mechanical flight control system, and FDD has become of primary interest. The state-of-practice, applied worldwide by all aircraft manufacturers, to diagnose these EFCS faults and obtain full flight envelope protection at all times is to provide high levels of hardware redundancy and dissimilarity in order to perform consistency tests and cross-checks. This also ensures sufficient available control action (fault tolerance). The interested reader can refer, among others, to [17–29].

2.2.3 Space Missions

For space missions, health monitoring is managed through a FDIR hierarchical approach in which several levels of faults are defined from local component/equipment up to global system failures [2, 7, 30]. Depending on the mission needs, FDIR functions are combined to other functions (data processing, orbitography, event-based commanding, and dynamic reprogramming) to achieve a desired level of availability, safety, and autonomy [2, 31–33]. FDIR strategy can be divided between all levels: detection and local reconfiguration in the subsystems, fault diagnosis and global reconfiguration at the operational level, and prevention at the decisional level (detect in advance plans that no longer consistent with the actual resource usage and may lead to further failures). The validation assumes testing all possible cross-path situations that becomes costly as the complexity of inboard hardware and software architectures increases. For early spacecraft, the above tasks were executed by sequential automata performing a priori known tasks. New-generation spacecraft has smart embedded systems, which are able to react to some known events and to select a decision among a predefined set. FDD, FTC, and FTG functions are strongly related to autonomy needs that vary with the mission scenarios and the expected benefits. Standardized degrees of autonomy can be found, for example, in [34]. See also [35] for an interesting discussion on autonomy needs for future space exploration missions. A low Earth orbit satellite can be endowed with an autonomous orbit control function to reduce ground operations. A deep space spacecraft, due to long communication delays, will require FDD and automatic reconfiguration capacities. For other space systems such as winged atmospheric reentry vehicles (e.g., space shuttle) which have aircraft-like configurations and more redundant control actuation, there are also more limited weight capabilities compounded because of more restrictive aerodynamic and controllability characteristics resulting from their lower lift-to-drag ratios. Note that, since the first flight of the Apollo mission where gain-variable Kalman filters were implemented into the Apollo lunar module first-generation digital flight computer, Kalman-based estimators are used in the flight control software of many space missions. Some estimated quantities could be used redundantly for fault detection and health management. The paper [36] describes the V&V challenges and approaches posed by the innovative FDIR technologies being employed and discusses additional certification considerations. The NASA technical report [37] discusses issues and lessons learned regarding designing, integrating, and implementing FDIR at Kennedy Space Center.

2.3 Review of Academic Advanced Results

2.3.1 Introduction

A large body of literature on FDD and FTC is now available. The open literature dealing with FTG is much more limited. Good surveys about academic state of the art can be found in [1, 3, 38–48]. FDD is a deep subject with hundreds of subtopics. The theory related to FDD has been developed since the early 1970s and can be considered today as a mature and well-structured field of research within the control community and offering many attractive features. FDD methods are classified generally into three categories, which include the knowledge or history-based methods [41, 49, 50], analytical model-based methods, and signal-based methods. For the latter, reference [44] gives a thorough review on the definitions and the methods for change detection with a main focus on the parametric statistical tools such as log-likelihood ratios and efficient scores. In this chapter, we will focus on analytical model-based approaches. Here by the term "model," we understand quantitative model: use of static and dynamic relations among system variables

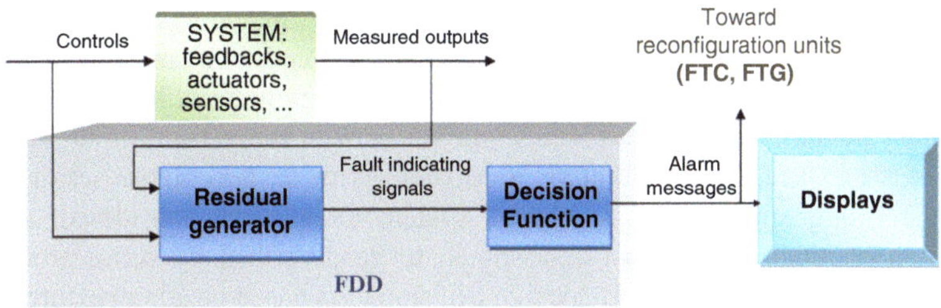

Fig. 2.6 Basic FDD structure

and parameters in order to describe system's behavior in quantitative mathematical terms. Note that the qualitative model-based methods, such as pattern recognition or rule-based approaches, capture discrepancies between observed behavior and that predicted by a qualitative model.

The early studies on model-based FDD appeared about 40 years ago. In [51–53], innovation signals are used to design detection filters. Many basic solutions have appeared during the 1980s: parity space and observer-based approaches, eigenvalue assignment, or parametric-based methods [1, 45–47, 54, 55]. In the 1990s, a great number of publications dealt with specific aspects such as robustness and sensitivity, diagnosis-oriented modeling, or robust isolation [38, 44, 47, 56–61]. The European school has been very active in the development of this field (see, e.g., and among others [38, 46–48, 62–71]). Today, and at least from a design point of view, model-based FDD can be considered as a mature field of research within the control community. The evidence of this can be seen through the very significant number of publications and dedicated international conferences.

2.3.2 *Analytical or Model-Based FDD*

The basic idea of model-based FDD is very simple and straightforward: residuals (fault indicating signals) are generated from comparison of the system measurements with their estimates. A threshold function (fixed or variable) can be used to provide additional levels of detection, while for fault isolation the generated residual has to include enough information to determine that a specific fault has occurred. The fault isolation is trivial in applications where the fault detector is dedicated to only one kind of fault.

The basic structure of a classical model-based FDD technique can be depicted as in Fig. 2.6.

The core element is the residual generation. Note that if only fault detection is of interest, reconstructing the fault rather than detecting its presence through a residual signal can be an alternative solution [64, 72–74]. Residual evaluation and decision making consist of checking the residuals and triggering alarm messages if the tolerances are exceeded. The thresholds can be set into different kinds. The simplest way is to use a constant threshold. The big advantage with fixed thresholds is their simplicity and reliability. Adaptive thresholds could enhance the sensitivity of fault detecting with the optimal choice of the magnitude which depends upon the nature of the system uncertainties and varies with the system input. Adaptive thresholds can keep the false alarm rate small with an acceptable sensitivity to faults. In some applications, stochastic system models are considered, and the generated residuals are known or assumed to be described by some probability distributions. It is then possible to design decision tests based on adaptive thresholds. More robust decision logics use the history of the residuals and utilize powerful or optimal statistical test techniques. The well-known examples of these statistical test techniques are sequential probability ratio test (SPRT), cumulative sum (CUSUM) algorithm, generalized likelihood ratio test, and local approach (see, e.g., [44]). To enhance the robustness of FDD schemes against small parameter variations and other disturbances during residual generation, different design, and evaluation tools have been proposed [38, 39]. The objective of any robust FDD method is to make the residuals become sensitive to one or more faults while at the same time making the residuals insensitive to modeling errors and uncertain disturbance effects acting upon the system being monitored. Robust FDD can be achieved if the residual signals maintain the desired sensitivity properties over a suitable range of the system's dynamic operation. A huge literature is now available dealing with various aspects of an FDD problem, ranging from modeling problems (nominal system modeling, fault modeling, disturbance and uncertainty modeling) and FDD system design.

The available design methods includes methods based on LTI, LPV, and nonlinear/hybrid estimators/observers, robust designs inspired by robust control designs, unknown input observers, and sliding-mode methods. The interested reader can refer, for example, to [38, 39] for recent surveys.

Remark 2.1 A hybrid system consists of a set of discrete modes, which represent fault states or operational modes of the system, and a set of continuous variables which model the continuous quantities that affect system behavior. Usually, the term state refers to the combination of these, that is, a state is a mode plus a value for each continuous variable, while the mode of a system refers only to the discrete part of the state.

Observer-based approaches have arisen as one of the most popular among FDI design techniques. In the linear case, it has been shown that any linear fault detection filter can be transformed into an equivalent observer-based form [75], providing a unified framework for analysis and implementation.

Generally speaking, the difficulties with LTI models for FDD lie in the need to produce meaningful models which can be used for synthesis and in the need of a posteriori robustness/sensitivity analysis. As it will be seen in Chap. 7, the effect of guidance, navigation, and control should be carefully analyzed and taken into account during model building. Modeling stage should also take into account uncertainties, stemming from a large variety of environmental disturbances and internal sources [76].

The things get much more complex in the nonlinear case from a design and also an analysis point of view. For a good recent survey on nonlinear FDD methods, the interested reader can refer to [48] and the references therein. Typically, the observer design problem is solvable if the system model can be transformed into a canonical form that may be a hard assumption to satisfy in many applications. An appealing approach to deal with some nonlinear problems is based on the LPV transformation. Consider, for example, a nonlinear system described by

$$\dot{x} = f(t, x, u, w), \; y = h(x) + v \tag{2.1}$$

where $x \in R^n$, $u \in R^m$, $w \in R^l$, $y \in R^p$, and $v \in R^p$ are, respectively, the state, the input, the disturbance, the output, and the measurement noise; $t \in R_+$ and the functions f, h are continuous with respect to all arguments and differentiable with respect to x and u. An LPV representation can be given by

$$\dot{x} = A(\rho(t))x + B(\rho(t))u, \; y = C(\rho(t))x + v \tag{2.2}$$

where the scheduling parameter vector $\rho \in \mathcal{P}$ is considered to be time varying (measured or estimated upon system operation) or unknown with known bounds; $\mathcal{P}$ is a set of functions that remain in a compact real subspace. The system (2.2) is an equivalent representation of (2.1), in the sense that all trajectories of (2.1) remain in the trajectories of (2.2). The basic idea is to replace nonlinear complexity of the model (2.1) by enlarged parametric variation in the linear model (2.2) which simplifies the design of an observer for (2.1). The main appeal of using the LPV formalism is that the solutions can be obtained using linear algebraic manipulations like those elaborated for LTI systems.

2.3.3 Recovery Aspects: FTC and FTG

The next step following the design of an FDD system is to decide appropriate recovery and corrective actions, based on all available actuator/sensor/communication resources. The recovery aspects have also been extensively studied (see, for instance, [3, 77]). The general objective is firstly to maintain stability and secondly to keep an acceptable performance level in fault situations. For successful reconfiguration actions, information about the failed element (fault identification) is necessary in order to access the remaining control resources. The interaction with the FDD system is a key point: generally FDD mechanism is supposed to detect and diagnose correctly any relevant signal degradation or failure. Obviously this must be done sufficiently early to set up timely recovery actions.

Usually the fault tolerance could be achieved through several potential solutions, for instance:

- Selecting a new precomputed control law depending on the faults which have been identified by the FDD system. In this case, hybrid control or switching control structures are commonly encountered in the literature [78].
- Synthesizing a new control strategy online. Such methods involve the calculation of new controller parameters once a failure has been identified by an online fault estimation scheme, following the typical design paradigm of adaptive control [79].
- Using dynamic control allocation for over actuated systems. The fault control allocation problem is that of distributing a desired total control effort among a redundant set of healthy actuators (without reconfiguration/accommodation of the controller) [80, 81].

The interested reader can refer to [82–87] and the references therein for further details.

The majority of the available methods rely implicitly on the assumption that the FDD and automatic reconfiguration and recovery systems are assumed to operate correctly, that is, the FDD outputs are supposed to be instantaneously available. The problem of guaranteeing stability and a certain level of performance of the overall fault-tolerant system, taking into account both the FDD performance (detection delay) and reconfiguration system, has not been sufficiently considered in the literature. Usually, the desired characteristics are checked a posteriori by means of Monte Carlo campaigns and nonlinear simulations. Note that for aerospace applications, validation assumes testing all possible cross-path situations that becomes costly with the GNC complexity increase and leads to intricate validation processes. Moreover, generally the sizing case corresponds to the worst performance that can be obtained in extreme situations. This procedure often limits the capability of "fail operational" strategies for some critical situations. Several more formal solutions have appeared recently. The effect of the FDD delay can be analyzed for linear systems [88]. In [89], a supervisory scheme uses a switching algorithm to fault isolation: a sequence of controllers is switched, until the appropriate one is found. Other works seek to combine a fault-tolerant controller and a diagnostic filter in both LTI and LPV settings (see, for instance, [82–85, 90]). However, the structure and parameters of the already in place control laws are generally modified. For aircraft systems, for example, this solution may lead to a new (long and expensive) certification campaign in fault-free situations. This could be a major concern for most safety-critical systems. Finally, FTG has been studied for some specific aerospace vehicles [4]. For example, for reusable launch vehicles (RLV), it has been shown in [91] that onboard autonomous FTG could be a promising solution, as it could provide a greater flexibility to account for off-nominal conditions or even to recover timely the vehicle from faulty situations.

2.4 Toward Advanced Model-Based Techniques for Flight Vehicles

2.4.1 Needs, Requirements, and Constraints

Aerospace industry needs continuous improvement including insertion of new technologies. Generally, new technologies are adopted in practice only when there is a clear cost or performance benefit. In aeronautics, at the same time, the main aircraft manufacturers tend more and more to use and adopt more sustainable technologies in order to decrease the environmental footprint of their airliners, feeding the needs for advanced strategies for accompanying any greener solutions. It should be noted that, from "a global air transport policy" point of view, much effort is being devoted to further improvement of sustainable and green air transport. The Single European Sky Air Traffic Management Research (SESAR) program in Europe and the Next Generation Air Transportation System (NextGen) in the USA seek to provide quicker flights, less fuel burn and emissions, shorter routes, and less congestion.

As an example, on the A380 airplane, the conventional hydraulic actuators have been replaced by a new generation of electrically powered actuators, the electro-hydrostatic actuator (EHA), mainly for reducing the number of hydraulic systems, generating significant weight and cost savings, and providing additional dissimilarity [92]. EHAs introduce new sources of faults that were tricky to detect with the state-of-practice FDD designs. However, any modification to the already proven and in-service solutions should undergo very long and stringent validation and verification process. Consider the example of a range checking fault detection method devoted to the detection of runaways in aircraft control surfaces servo-loops [93]. This simple technique provides sufficient fault coverage and ensures a perfect robustness without false alarm. The choice of any other "advanced" candidate solution should be clearly demonstrated in terms of added value from an industrial point of view. This means that any changes to existing and already proven scheme should provide a viable technological solution ensuring either better performance while guaranteeing the same level of robustness, or better robustness for the same level of performance, or better performance and better robustness and covering larger fault profile. More generally, the selection of an advanced solution at a local or global level for aerospace missions necessarily includes a trade-off between the best adequacy of the technique and its implementation level for covering an expected fault profile. For proper implementation, those techniques should be embedded within the physical redundancy structure of the system. New methods and technologies entail risk and thus, despite potential cost, performance, and sustainability advantages, must undergo extensive development, validation, and verification before they can be transitioned to real-world systems. That is why decision makers, by default, rely on already proven technical solutions. This is especially true for space applications, as solutions cannot be tested beforehand due to the difficulties of reproducing space-representative conditions on Earth. For space missions, there exist however a number of challenging requirements to meet the autonomy needs of future space missions. Examples of which are Mars exploration missions and the in-the-drawing-board science missions involving multi-craft formation flying, Near-Earth Objects (NEO), or deep space exploration in general. For space systems, the usual implementation constraints found in aeronautics, such as computation load and complexity, are also encountered albeit to a greater degree due to the more limited weight and computational processing capabilities. These more restrictive limitations arise from the expensive cost for putting additional payload in space and by the lengthier testing and validation process required to classify any design as space ready. The weight limitation directly affects the system decisions related to hardware redundancy, while the processing limitation affects those decisions related to the choice of the onboard diagnosis capabilities and reconfiguration techniques. In the civil aircraft industry, compared to space missions, not only one model is manufactured but hundreds of aircraft are generally mass-produced during several decades. All along the aircraft production, some modifications can be envisaged: extended range, increased maximum take-off weight, extended passenger capacity, etc. In this context, one crucial requirement is the adaptation of the new methods to slightly different aircraft models. For example, a given FDD technique cannot be tuned on a case-by-case basis, but must be generic enough for different versions of the aircraft. In the same order of idea, another requirement concerns the adaptability of the design from one system to the other or even from one control surface to the other. Suppose, for example, that a given FDD technique has been developed for one inboard ailerons of the A380. This FDD technique may be called to be used on the outboard ailerons. If the FDD system requires a completely different tuning of the design via complex methods, it will be difficult to be mastered by the development teams and will penalize the transition to the industrial world. Easy-to-tune high-level input parameters are necessary for the adaptability of a new solution in the framework of mass-production. A limited number of tuning parameters is also desirable for shortening the validation and verification activities demanded for certification. These aspects will be discussed more in details in Chaps. 3 and 4.

2.4.2 *Case Studies*

One can find a lot of "case study" in the open literature which is fragmented across many technical papers. See, for example, and among others, [4, 62, 94–108], and many technical reports available at http://www.sti.nasa.gov/.

For space missions, one can mention the precursor NASA's New Millennium Program [109]: here, the so-called Deep Space One (DS1) Remote Agent Experiment was initiated to demonstrate onboard fault-protection capabilities, including failure diagnosis and recovery, onboard replanning following otherwise unrecoverable failures, and system-level fault protection [110]. Another example is L2 (Livingstone2) program [111] which flew on the Deep Space One spacecraft as part of the Remote Agent Experiment in May 1999. In Livingstone, diagnosis is done by maintaining a candidate hypothesis (in other systems more than one hypothesis is kept) about the current state of each system component and comparing the candidate's predicted behavior with the system sensors. Analytical redundancy and Bayesian decision theory were combined to produce a sensor validation system concept for real-time monitoring of Space Shuttle Main Engine telemetry [112]. The validation system was implemented in Ada and hosted on a Boeing X-33 prototype flight computer. In [36], the authors present a work related to the certification of a pilot application of advanced FDIR software at Ames Research Center and at the Jet Propulsion Laboratory (NASA). The authors underline the stringent requirements in terms of test effort and the value of rethinking V&V when novel technologies are being deployed.

In the open literature, there exist a great number of studies dealing with FDD and FTC in flight control systems (see, e.g., [4, 8, 70, 97, 98, 108, 113, 114]). In Europe, the FDD challenges for aircraft flight control systems were investigated within the ADDSAFE project [115]. Here, by introducing advanced FDD techniques, the goal was to contribute to achieve the European Vision 2020 challenges related to the "greening" of the aircraft. Analytical redundancy has been used on A380 aircraft for the detection of a specific failure case related to EFCS [116].

Insertion of new technologies is assessed by TRL measure [117]. TRL provides a significant input to risk assessment of including a technology in an existing or new program. Roughly speaking, academic activities cover TRL1 (basic principles) up to TRL3 (laboratory and case studies, validation on high-fidelity simulators). TRL6 (prototype demonstration) – TRL9 ("flight proven" through successful mission operations) correspond to technology integration. Often, despite clear needs, new technologies require several years of maturation to the point of practical usefulness, i.e., reaching high TRL. That is why we can observe a "Death Valley" corresponding to TRL4 – TRL5 (validation in relevant environment). This applicability gap has resulted in a real technological barrier. A number of ongoing works at NASA are devoted to bridge this gap. None of the above mentioned remarks are intended to minimize the importance of academic developments. However, it is important to recognize that the gap on the whole is large and warrants serious introspection by the research community. Bridging the gap, from the researcher's perspective, requires that new methods and techniques be communicated to engineers who are in a position to apply them. Motivations which are behind new academic developments should be presented in a more practically relevant way. As an example, many of the early published academic papers on model-based FDD start with the statements such as "hardware redundancy is expensive, heavy, less potentially reliable, it should be replaced by model-based techniques whereby additional knowledge of the system is leveraged instead of actual redundancy...." This basic and historical argument which played a driving role to motivate the early development of FDD academic research could be rather misleading when applied to the aerospace vehicles. A good balance between conventional, technically proven and in-service solutions, and advanced model-based techniques is probably the only right solution in many applications. This observation has been pointed out in [6] where the author developed several interesting ideas about redundancy management. Model-based techniques do not substitute for physical redundancy but it can be a useful and powerful supplement, if implemented in a manner that properly exploits the physical redundancy.

2.5 Conclusions

There is a growing need to move toward greater onboard reconfiguration capacities and earlier robust diagnosis of system malfunctions. For space missions, this need is driven by the more challenging requirements for future space missions under limited weight and computational processing capabilities. For new-generation civilian aircraft, the need is driven by the more and more stringent requirements which would come in force for future and more environmentally friendlier programs.

The basic aim of this chapter was to give an overview of various model-based approaches to FDD and automatic reconfiguration and the state-of-the-art efforts in terms of industrial applications for aerospace systems. The picture is certainly not complete because of the huge number of various works and studies available in the literature. The focus was to show that while research went forward since the early 1970s, the design methodology involving feasibility analysis and real-world requirements specification is still missing, despite efforts in the past few years. Important issues are potential reduction of physical redundancy, overall reliability, robustness in harsh environments and worst-case performance evaluation. These issues will be discussed in the following chapters through a number of aerospace applications.

References

1. Isermann R (1997) Trends in the application of model-based fault detection and diagnosis of technical processes. Control Eng Pract 5(5):709–719
2. Olive X (2012) FDI(R) for satellites: how to deal with high availability and robustness in the space domain. Int J Appl Math Comput Sci 22(1):99–107. doi: 10.2478/v10006-012-0007-8
3. Blanke M, Kinnaert M, Lunze M, Staroswiecki M (2003) Diagnosis and fault tolerant control. Springer, New York
4. Ducard GJJ (2009) Fault-tolerant flight control and guidance systems, Advances in industrial control. Springer, London
5. Noura H, Theilliol D, Ponsart J-C, Chamseddine A (2009) Fault-tolerant control systems. Design and practical applications, Advances in industrial control. Springer, London
6. Osder S (1999) Practical view of redundancy management, application and theory. J Guid Control Dyn 22(1):12–21
7. Tomayko JE (2000) Computers take flight: A history of NASA's pioneering digital fly-by-wire project. NASA-SP-2000-4224. Available at http://www.nasa.gov/centers/dryden/pdf/182985main_DFBW_rev1.pdf

8. Philippe C et al (2011) Aerospace control. In: Samad (Honeywell) T, Annaswamy (MIT) A (eds) The impact of control technology, overview, success stories, and research challenges. IEEE Control System Society. Available at: http://ieeecss.org/general/impact-control-technology
9. Goupil P (2011) AIRBUS state of the art and practices on FDI and FTC in flight control system. Control Eng Pract 19:524–539
10. Palmer MT, Abbot KH (1994) Effects of expected-value information and display format on recognition of aircraft subsystem abnormalities. NASA TP-3395, March 1994, A technical paper 3395
11. Regal DM, Rogers VH, Boucek GP (1989) Situational awareness in the commercial flight deck – definition, measurement, and enhancement. In: Proceedings of the 7th aerospace behavioral technology conference and exposition. SAE, Warrendale, pp 65–69
12. Johnson DM (1996) A review of fault management techniques used in safety-critical avionic systems. Prog Aerosp Sci 32(5):415–431(17)
13. Trujillo AC (1998) Pilot mental workload with predictive system status information. In: 4th annual symposium on human interaction with complex systems, Fairborn, OH, pp 73–80
14. Favre C (1994) Fly-by-wire for commercial aircraft: the Airbus experience. Int J Control 59(1):139–157
15. Traverse P, Lacaze I, Souyris J (2004) Airbus fly-by-wire: a total approach to dependability. In: Proceedings of the 18th IFIP world computer congress, Toulouse, pp 191–212
16. FAR/CS 25, Airworthiness standards: transport category airplane, published by FAA, title 14, part 25, and certification specifications for large aeroplanes, published by EASA, CS-25
17. Alcorta-Garcia E, Zolghadri A, Goupil P (2011) Nonlinear observer-based strategy for aircraft oscillatory failure detection: A380 case study. IEEE Trans Aerosp Electron Syst 47:2792–2806
18. Briere D, Traverse P (1993) Airbus A320/A330/A340 electrical flight controls—a family of fault-tolerant systems. In: Proceedings of the 23rd international symposium on fault-tolerant computing, Toulouse, pp 616–623
19. Chen RH, Ng HK, Speyer JL, Guntur LS, Carpenter R (2004) Health monitoring of a satellite system. In: Proceedings of AIAA guidance, navigation, and control conference, Minneapolis, August 2004
20. Kumar M (2007) Fault detection identification and reconfiguration of flight control system using IMM estimator. In: Proceedings of the digital avionics systems conference, October 2007
21. Jung B, Kim Y, Ha C, Tahk MJ (2007) Nonlinear reconfigurable flight control system using multiple model adaptive control. Presented at the 17th IFAC symposium on automatic control aerospace, Toulouse, France, June 2007
22. Tang XD, Tao G, Joshi SM (2003) Adaptive actuator failure compensation for parametric strict feedback systems and an aircraft application. Automatica 39(11):1975–1982
23. Oppenheimer MW, Doman DB (2006) Efficient reconfiguration and recovery from damage for air vehicles. Presented at the AIAA guidance, navigation, and control conference, Keystone, CO, August 2006
24. Ganguli G, Papageorgiou, Glavaski S (2006) Aircraft fault detection, isolation and reconfiguration in the presence of measurement errors. Presented at the AIAA guidance, navigation, and control conference, Keystone, CO, August 2006
25. Cieslak J, Henry D, Zolghadri A (2010) Fault tolerant flight control: from theory to piloted flight simulator experiments. IET Control Theory Appl 4:1451–1464
26. Cieslak J, Henry D, Zolghadri A, Goupil P (2008) Development of an active fault tolerant flight control strategy. AIAA J Guid Control Dyn 31:135–147
27. Edwards C et al (2010) Fault tolerant flight control – a benchmark challenge. Lecture Notes in Control and Information Sciences
28. Lopez I, Sarigul-Klijn N (2010) A review of uncertainty in flight structural damage monitoring, diagnosis and control. Prog Aerosp Sci 46:247–273

29. Gheorghe A, Zolghadri A, Cieslak J, Goupil P, Dayre R, Le Berre H (2013) Toward model-based approaches for fast and robust fault detection in aircraft control surface servo-loop: from theory to application. IEEE Control Syst Mag, June 2013
30. Butler RW (2008) A primer on architectural level fault tolerance. NASA/TM-2008-215108. Langley Research Center, Hampton, VA
31. Lemai S, OLive X, Charmeau MC (2006) Decisional architecture for autonomous space systems. In: 9th ESA workshop on advanced technologies for robotics and automation, Noordwijk, The Netherlands, 28–30 Nov 2006
32. Durou O, Godet V, Mangane L, Perarnaud DP, Roques R (2002) Hierarchical fault detection, isolation and recovery applied to COF and ATV avionics. Acta Astronaut 50(9):547–556
33. Ferrell B, Lewis M, Perotti J, Oostdyk R, Brown B (2010) Functional fault modeling conventions and practices for real-time fault isolation. Ames Research Center; Kennedy Space Center. Available at http://ntrs.nasa.gov/search.jsp?R=20110004336
34. ECSS 70-11A (2005) Space engineering: space segment operability. European Cooperation for Space Standardization standard, August 2005
35. Truszkowski WF, Hinchey MG, Rash JL, Rouff CA (2006) Autonomous and autonomic systems: a paradigm for future space exploration missions. IEEE Trans Syst Man Cybern Part C Appl Rev 36(3):279–291
36. Feather MS, Markosian LZ (2008) Towards certification of a space system application of fault detection and isolation. In: International conference on prognostics and health management, Denver, CO, October 2008
37. Ferell B, Lewis M, Perotti J, Oostdyk R, Goerz J, Brown R (2010) Lessons learned on implementing fault detection, isolation, and recovery (FDIR) in a ground launch environment. Ames Research Center; Kennedy Space Center. Technical report available at http://ntrs.nasa.gov/search.jsp?R=20110004130
38. Ding SX (2008) Model-based fault diagnosis techniques: design schemes, algorithms, and tools. Springer, Berlin/Heidelberg
39. Hwang I, Kim S, Kim Y (2010) A survey on fault detection, isolation and reconfiguration methods. IEEE Trans Control Syst Technol 18(3):636–653
40. Venkatasubramanian V, Rengaswamy R, Yin K, Kavuri SN (2003) A review of process fault detection and diagnosis Part I: Quantitative model-based methods. Comput Chem Eng 27:293–311
41. Venkatasubramanian V, Rengaswamy R, Yin K, Kavuri SN (2003) A review of process fault detection and diagnosis Part II: Qualitative models and search strategies. Comput Chem Eng 27:313–326
42. Venkatasubramanian V, Rengaswamy R, Yin K, Kavuri SN (2003) A review of process fault detection and diagnosis Part III: Process history based methods. Comput Chem Eng 27: 327–346
43. Isermann R (2005) Model-based fault-detection and diagnosis status and applications. Annu Rev Control 29(1):71–85
44. Basseville M, Nikiforov IV (1993) Detection of abrupt changes: theory and application. Prentice Hall, Englewood Cliffs
45. Patton R, Frank PM, Clark RN (1989) Fault diagnosis in dynamic systems: theory and application. Prentice-Hall, Englewood Cliffs
46. Patton R (1997) Fault-tolerant control: the 1997 situation. In: SAFEPROCESS'97, IFAC Symposium on fault detection, supervision and safety, Kingston Upon Hull, UK
47. Chen J, Patton RJ (1999) Robust model-based fault diagnosis for dynamic systems. Kluwer Academic, Boston/Dordrecht/London
48. Bokor J, Szabo Z (2009) Fault detection and isolation in nonlinear systems. Annu Rev Control 33:113–123
49. Cordier MO, Dague P, Lévy F, Mountmain J, Staroswiecki M, Travé-Massuyès L (2004) Conflicts versus analytical redundancy relations: a comparative analysis of the model based diagnosis approach from the artificial intelligence and automatic control perspectives. IEEE Trans Syst Man Cybern B Cybern 34(5):2163–2177

50. Travé-Massuyès L, Escobet T, Olive X (2006) Diagnosability analysis based on component-supported analytical redundancy relations. IEEE Trans Syst Man Cybern A Syst Hum 36(6):1146–1160
51. Beard RV (1971) Failure accommodation in linear systems through self-reorganization. PhD dissertation, Department of Aeronautics Astronautics, Massachusetts Institute of Technology, Cambridge
52. Jones HL (1973) Failure detection in linear systems. PhD dissertation. Department of Aeronautics Astronautics, Massachusetts Institute of Technology, Cambridge, MA
53. Mehra RK, Peschon J (1971) An innovations approach to fault detection and diagnosis in dynamic systems. Automatica 7:637–640
54. Massoumnia MA (1986) A geometric approach to the synthesis of failure detection filters. IEEE Trans Autom Control 31(9):839–846
55. Chow EY, Willsky AS (1984) Analytical redundancy and the design of robust failure detection systems. IEEE Trans Autom Control 29(7):603–614
56. Zolghadri A, Goetz C, Bergeon B, Denoise X (1998) Integrity monitoring of flight parameters using analytical redundancy. In: Proceedings of the UKACC international conference on control (CONTROL '98), Swansea, UK, pp 1534–1539
57. Zolghadri A (1996) An algorithm for real-time failure detection in Kalman filters. IEEE Trans Autom Control 41(10):1537–1540
58. Douglas RK, Speyer JL (1996) Robust fault detection filter design. J Guid Control Dyn 19(1):214–218
59. Chen J, Patton RJ, Zhang HY (1996) Design of unknown input observers and robust fault-detection filters. Int J Control 63(1):85–105
60. Balas MJ (1999) Do all linear flexible structures have convergent second order observers? AIAA J Guid Control Dyn 22(6):905–908
61. Stoustrup J, Niemann JH (2002) Fault estimation—a standard problem. Int J Robust Nonlinear Control 12:649–673
62. Zolghadri A, Castang F, Henry D (2006) Design of robust fault detection filters for multivariable feedback systems. Int J Model Simul 26:17–26
63. Bokor J, Balas G (2004) Detection filter design for LPV systems – a geometric approach. Automatica 40:511–518
64. Yan XG, Edwards C (2007) Nonlinear robust fault reconstruction and estimation using a sliding mode observer. Automatica 43:1605–1614
65. Henry D, Zolghadri A (2005) Design and analysis of robust residual generators for systems under feedback control. Automatica 41:251–264
66. Henry D, Zolghadri A (2005) Design of fault diagnosis filters: a multi-objective approach. J Franklin Inst 342(4):421–446
67. Henry D, Zolghadri A (2006) Norm-based design of robust FDI schemes for uncertain systems under feedback control: comparison of two approaches. Control Eng Pract 14(9):1081–1097
68. Zolghadri A, Henry D, Grenaille S (2008) Fault diagnosis for LPV systems. In: 16th IEEE Mediterranean conference on control and automation, Ajaccio, France
69. Grenaille S, Henry D, Zolghadri A (2008) A method for designing fault diagnosis filters for LPV polytopic systems. J Control Sci Eng. Article ID 231697, Vol 1, January 2008
70. Ganguli S, Marcos A, Balas G (2002) Reconfigurable LPV control design for Boeing 747-100/200 longitudinal axis. In: American control conference. Anchorage, Alaska, USA, 8–10 May 2002
71. Henry D, Zolghadri A, Monsion M, Ygorra S (2002) Off-line robust fault diagnosis using the generalized structured singular values. Automatica 38:1347–1358
72. Raissi T, Videau G, Zolghadri A (2010) Interval observers design for consistency checks of nonlinear continuous-time systems. Automatica 46:518–527
73. Efimov D, Zolghadri A, Raissi T (2011) Actuators fault detection and compensation under feedback control. Automatica 47:1699–1705

74. Saif M, Xiong Y (2003) Sliding mode observers and their application in fault diagnosis. In: Caccavale F, Villani L (eds) Fault diagnosis and fault tolerance for mechatronic systems: recent advances, vol 1, Springer tracts in advanced robotics. Springer, Berlin, pp 1–57
75. Alazard D, Apkarian P (1999) Exact observer-based structures for arbitrary compensators. Int J Robust NL Control 9:101–118
76. Zhou K, Doyle JC, Glover K (1995) Robust and optimal control. Prentice Hall, Upper Saddle River
77. Zhang Y, Jiang J (2008) Bibliographical review on reconfigurable fault-tolerant control systems. Ann Rev Control 32(2):229–252
78. Yang H, Cocquempot V, Jiang B (2009) Robust fault tolerant tracking control with applications to hybrid nonlinear systems. IET Control Theory Appl 3(2):211–224
79. Staroswiecki M, Yang H, Jiang B (2007) Progressive accommodation of parametric faults in LQ control. Automatica 43:2070–2076
80. Alwi H, Edwards C (2008) Fault tolerant control using sliding modes with on-line control allocation. Automatica 44:1859–1866
81. Hamayun MT, Edwards C, Alwi H (2010) Integral sliding mode fault tolerant control incorporating on-line control allocation. In: 11th international workshop on variable structure systems, Mexico City, 26–28 June 2010
82. Ding SX (2009) Integrated design of feedback controllers and fault detectors. Annu Rev Control 33:124–135
83. Gaspar P, Bokor J (2006) A fault-tolerant rollover prevention system based on a LPV method. Int J Veh Des 42(3–4):392–412
84. Liberzon D (2003) Switching in systems and control. Birkhäuser, Boston
85. Marcos A, Balas G (2005) A robust integrated controller/diagnosis aircraft application. Int J Robust NL Control 15:531–551
86. Weng Z, Patton R, Cui P (2008) Integrated design of robust controller and fault estimator for linear parameter varying systems. In: 17th World Congress IFAC, Seoul, Korea
87. Zhang Y, Jiang J (2008) Bibliographical review on reconfigurable fault-tolerant control systems. Annu Rev Control 32:229–252
88. Shin J-Y, Belcastro CM (2006) Performance analysis on fault tolerant control system. IEEE Trans Control Syst Technol 14(9):1283–1294
89. Yang H, Jiang B, Staroswiecki M (2009) Supervisory fault tolerant control for a class of uncertain nonlinear systems. Automatica 45:2319–2324
90. Oudghiri M, Chadli M, El Hajjaji A (2008) Robust observer-based fault tolerant control for vehicle lateral dynamics. Int J Veh Des 48:173–189
91. Morio V (2009) Contribution au développement d'une loi de guidage autonome par platitude. Application à une mission de rentrée atmosphérique (Shuttle orbiter STS-1). PhD dissertation, Bordeaux 1 university, Bordeaux, France
92. Van den Bossche D (2006) The A380 flight control electrohydrostatic actuators, achievements and lessons learnt. In: Proceedings of the 25th Congress of the International Council of the Aeronautical Sciences, Hamburg, Germany
93. Zolghadri A, Gheorghe A, Cieslak J, Henry D, Goupil P, Dayre R, Le Berre H (2011) A model-based solution to robust and early detection of control surface runaways. SAE Int J Aerosp 4:1500–1505
94. Kurtoglu T, Johnson SB, Barszcz E, Johnson JR, Robinson PI (2008) Integrating system health management into the early design of aerospace systems using Functional Fault Analysis. IEEE conference on prognostics and health management, Denver
95. Deckert JC, Desai MN, Deyst JJ, Willsky AS (1977) F-8 DFBW sensor failure identification using analytic redundancy. IEEE Trans Autom Control 22(5):795–809
96. Wilbers DM, Speyer JL (2002) Detection filters for aircraft sensor and actuator faults. In: Proceedings of the IEEE international conference on control applications, Jerusalem, Israel
97. Menke TE, Maybeck PS (1995) Sensor/actuator failure detection in the VISTA F-16 by multiple model adaptive estimation. IEEE Trans Aerosp Electron Syst 31(4):1218–1229

98. Kim S, Choi J, Kim Y (2008) Fault detection and diagnosis of aircraft actuators using fuzzy-tuning IMM filter. IEEE Trans Aerosp Electron Syst 44(3):940–952
99. Chen RH, Ng HK, Speyer JL, Guntur LS, Carpenter R (2004) Health monitoring of a satellite system. In: Proceedings of the AIAA guidance, navigation, and control conference, Boston, August 2004
100. Patton R, Uppal F, Simani S, Polle B (2010) Robust FDI applied to thruster faults of a satellite system. Control Eng Pract 18(9):1093–1109
101. Falcoz A, Henry D, Zolghadri A (2010) Robust fault diagnosis for atmospheric re-entry vehicles: a case study. IEEE Trans Syst Man Cybern Part A Syst Hum 40:886–899
102. Falcoz A, Henry D, Zolghadri A, Bornschleg E, Ganet M (2008) On-board model-based robust FDIR strategy for reusable launch vehicles (RLV). In: 7th international ESA conference on guidance, navigation and control systems, County Kerry, Ireland
103. Henry D, Falcoz A, Zolghadri A (2009) Structured $H\infty/H-$ LPV filters for fault diagnosis: some new results. In: 7th IFAC symposium on fault detection, supervision and safety of technical processes, Barcelona, Spain
104. Henry D (2008) Fault diagnosis of the microscope satellite actuators using Hinf/H- filters. AIAA J Guid Control Dyn 31(3):699–711
105. Zolghadri A (2000) A redundancy-based strategy for safety management in a modern civil aircraft. Control Eng Pract 8(5):545–554
106. Zolghadri A (2002) Early warning and prediction of flight parameter abnormalities for improved system safety assessment. Reliab Eng Syst Saf 16:19–27
107. Papageorgiou C, Glover K (2005) Robustness analysis of nonlinear flight controllers. AIAA J Guid Control Dyn 28(4):639–648
108. Rotstein HP, Ingvalson R, Keviczky T, Balas GJ (2006) Fault-detection design for uninhabited aerial vehicles. J Guid Control Dyn 29(5):1051–1060
109. James M, Dubon L (2000) An autonomous diagnostic and prognostic monitoring system for NASA's deep space network. Proc IEEE Aerosp Conf 2:403–414
110. Bernard D, Dorais G, Gamble E, Kanefsky B, Kurien J, Man G, Millar W, Muscettola N, Nayak P, Rajan K, Rouquette N, Smith B, Taylor W, Tung YW (1999) Spacecraft autonomy flight experience: the DS1 remote agent experiment. In: Proceedings of AIAA, Albuquerque, NM, pp 28–30
111. http://www.nasa.gov/centers/ames/research/technology-onepagers/livingstone2-modelbased.html
112. Bickford RL et al (1999) Real-time sensor data validation for space shuttle main engine telemetry monitoring. In: AIAA/ASME/SAE/ASEE 35th joint propulsion conference and exhibit, Los Angeles, CA
113. Lombaerts TJJ (2010) Fault tolerant flight control. A physical model approach. PhD thesis, TuDelft
114. Azam M, Pattipati K, Allanach J, Poll S, Patterson-Hine A (2005) In-flight fault detection and isolation in aircraft flight control systems. In: IEEE aerospace conference, 5–12 Mar 2005
115. http://addsafe.deimos-space.com
116. Goupil P (2010) Oscillatory failure case detection in the A380 electrical flight control system by analytical redundancy. Control Eng Pract 18(9):1110–1119
117. http://en.wikipedia.org/wiki/Technology_readiness_level

Chapter 3
Robust Detection of Oscillatory Failure Case in Aircraft Control Surface Servo-Loops

Acronyms

EFCS	Electrical Flight Control System
FCC	Flight Control Computer
FDD	Fault Detection and Diagnosis
OFC	Oscillatory Failure Case
AAB	Airbus Aircraft Benchmark
FES	Functional Engineering Simulator
ATF	Airbus Test Facilities
FOM	Figures-of-Merits
DTP	Detection Time Performance
FA	False Alarm
MD	Missed Detection
ET	Executive Time
V&V	Validation and Verification

3.1 Introduction and Motivations

3.1.1 Primary Aircraft Control Surfaces

In addition to thrust control, the principal means of controlling an aircraft is through aerodynamic forces generated by control surfaces which are generally movable flaps located on the fuselage, wing, and tail. The primary purpose of certain control surfaces (e.g., elevator, rudder, and ailerons) is to generate control moments; hence, their resultant forces act at some distance from the aircraft center of mass. In this section the main control surfaces and their functions are briefly recalled (see Fig. 3.1).

A. Zolghadri et al., *Fault Diagnosis and Fault-Tolerant Control and Guidance for Aerospace Vehicles: From Theory to Application*, Advances in Industrial Control,
DOI 10.1007/978-1-4471-5313-9_3, © Springer-Verlag London 2014

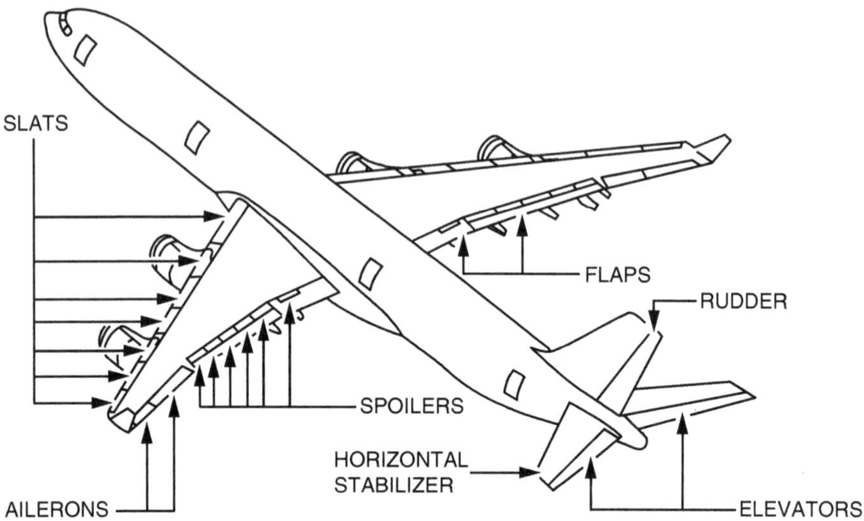

Fig. 3.1 Primary and secondary flight control surfaces on an Airbus A340

Let us take the example of a typical Airbus civil aircraft to illustrate what is the flight control system (FCS) and its main components, among which the aforementioned control surfaces. In manual control mode, the FCS consists of all the elements located between the pilot inputs (in the cockpit) and the control surfaces, including these two elements. This includes also sensors, probes, actuators, power sources (hydraulic and electrical), wiring, and flight control computers. In automatic mode, the FCS is coupled to the autopilot system. The flight control surfaces are comprised of the rudder, the elevators, the Trimmable Horizontal Stabilizer (THS), the ailerons, the spoilers, and the slats and flaps (see Fig. 3.1). Excluding the slats and flaps, the FCS is generally termed as the primary FCS. The slats and flaps associated to all their control and monitoring devices are called the secondary FCS and are used to control the aircraft lift. The primary FCS is intended to control the aircraft attitude, trajectory, and speed in manual mode. Its main objectives also include: flight safety, reduction of the pilot workload, fault-tolerant control (including fault detection), automatic reconfiguration after fault detection, optimization of the aircraft performances, and passenger comfort.

The elevators are used to control, in the short term, the longitudinal movement of the aircraft. The THS allows for a long-term longitudinal maneuver and enables to keep an average deflection of the elevators around zero despite center of gravity position changes, speed fluctuation, and aerodynamic configuration variations. This allows for decreasing drag and for keeping the maximum possible elevator deflection. To sum up, the THS allows the aircraft to be balanced, and the elevators allow the aircraft to be controlled around a steady position. The ailerons are used in anti-symmetrical way to create a roll moment (aircraft turn). The spoilers are utilized either symmetrically to break the lift (ground spoilers, emergency descent) or anti-symmetrically to contribute to the roll moment, if necessary. The rudder is used to create a yaw moment mainly for controlling the position of the nose of the aircraft but also for yaw damping and turn coordination.

It should be pointed out that a particular element (e.g., the wing) produces a primary effect (lift), and, at the same time, it may also produce secondary effects (drag and side forces, as well as pitching, yawing, and rolling moments). Because the components are in close proximity, the aerodynamic forces and moments that they generate are interrelated. In most cases, it is desirable for a control surface deflection to produce a single control force or moment proportional to the deflection. The constants of proportionality change with dynamic pressure, Mach number, and angle of attack, i.e., they are affected by flow interference and airframe elasticity.

The interested reader can refer to [1] for a comprehensive description of airframe components and flight dynamics.

3.1.2 The Link Between FDD of Control Surfaces and Aircraft Structural Design

Regulations (for instance, CS 25.302; see reference FAR/JAR 25) used for aircraft certification state that the system must be designed so that it cannot produce unexpected high loads on the aircraft. However, some EFCS-failure cases can influence structural loads, for example, loss of limitations (e.g., rudder deflection limitation in function of aircraft speed), loss of an EFCS special function to reduce structural design loads (e.g., load alleviation function), or degradation of deflection rates. The capability to detect such failures is of primary interest because it has an impact on the structural design of the aircraft. The failure amplitude must be contained by system design within a given envelope. Dedicated monitoring must be used to guarantee that the failure will remain within an envelope with acceptable robustness. In other words, if a failure of a given amplitude cannot be detected and "passivated" (which means that the propagation of the failure is stopped), this amplitude must then be considered for load computations. The result of this computation can lead to reinforce the structure. Robust and earlier FDD of such faults with smaller magnitude allows the designers to avoid reinforcing the structure and to save weight in order to help aircraft achieve sustainability goals (fuel burn, noise, range, and environmental footprint).

Historically, the EFCS or flight-by-wire (FBW) was already a weight-saving technology since the conventional mechanical linkages between the pilot's stick and the control surface actuators are replaced by electrical signal wires. More recently,

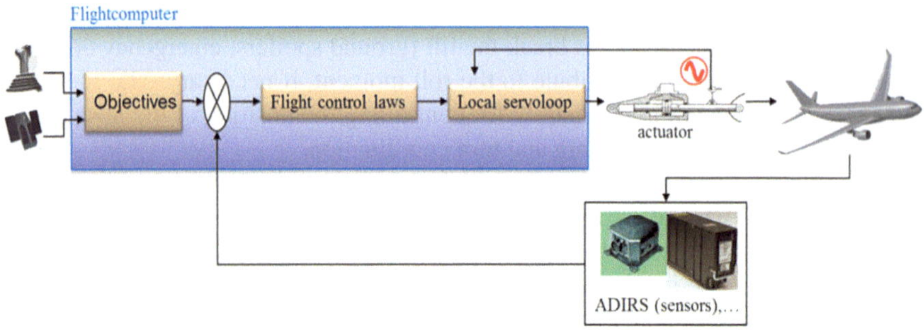

Fig. 3.2 Propagation of an oscillatory failure

the introduction of Electro-Hydraulic Actuators (EHA) on the A380 airplane [2] allowed for replacing the three conventional hydraulic circuitries by two hydraulic ones plus two electric layouts, which saves around 1 ton mass for the aircraft.

Typical EFCS-failure cases causing significant structural loads also include runaway, jamming, and oscillation of control surfaces. The latter is called Oscillatory Failure Cases (OFC). The objective of this chapter and the following is to propose model-based techniques to detect such events in an early stage. This chapter will deal with OFC in EFCS.

The presented techniques have been tested and validated on an industrial benchmark, Airbus Test Facilities, and Airbus A380 flight simulator.

3.1.3 Oscillatory Failure Case

Oscillatory Failure Case (OFC) is an abnormal oscillation of a control surface due to component malfunction in control surface servo-loops. This signal, of unknown amplitude and frequency, can be propagated downstream the control loop to the control surface and could excite the airplane structure producing structural loads (Fig. 3.2) [2]. If OFCs of given amplitude cannot be detected and passivated in time, this amplitude must be considered for load computations. If the result of this computation falls outside the load envelope, then it is necessary to reinforce the structure. So, in order to avoid reinforcing the structure and consequently to save weight, low amplitude OFCs must be detectable at a very early stage. This is an important feature in the context of aircraft overall design optimization and for supporting the development of a more sustainable aircraft. Note that because OFC is of unknown amplitude and frequency, many classical methods cannot be applied for its reliable detection [3]. Consequently, robust and early detection of OFC appears to be a challenging problem.

3.2 OFC in Aircraft Control Surface Servo-Loop

3.2.1 Description

OFCs arise mainly due to electronic components or to actuator mechanical parts in faulty mode generating spurious harmonic signals. This oscillatory signal propagates through the servo-loop control, leading to an unwanted control surface oscillation. The faulty components can be located inside the flight control computer (FCC) analog inputs/outputs, the actuator position sensors, or the actuators. The FCC may also generate unwanted oscillations of the command current sent to the actuator servo-valve.

As the considered faults are located in the servo-control loop of the moving surfaces, between the FCC and the control surfaces, including these two elements, the faults impact only one control surface. Potential locations of the OFC source are shown in Fig. 3.3.

OFC may appear as a so-called liquid or solid failure at the control surface level [2]. Liquid OFC are additive faults: they are added to the healthy signal, and thus the control surface deflects according to the superimposition as propagated by EFCS. A solid OFC substitutes the nominal signal, and the control surface executes a pure periodic motion. Note that in this case it is not possible to correct the faulty signal as the control signal has no more effect (it is replaced by the spurious solid fault signal). OFCs are considered as harmonic signals with frequency and amplitude uniformly distributed generally over the frequency range 0.1–10 Hz. Beyond 10 Hz, OFCs have no significant effects because of the low-pass behavior of the actuator. The time

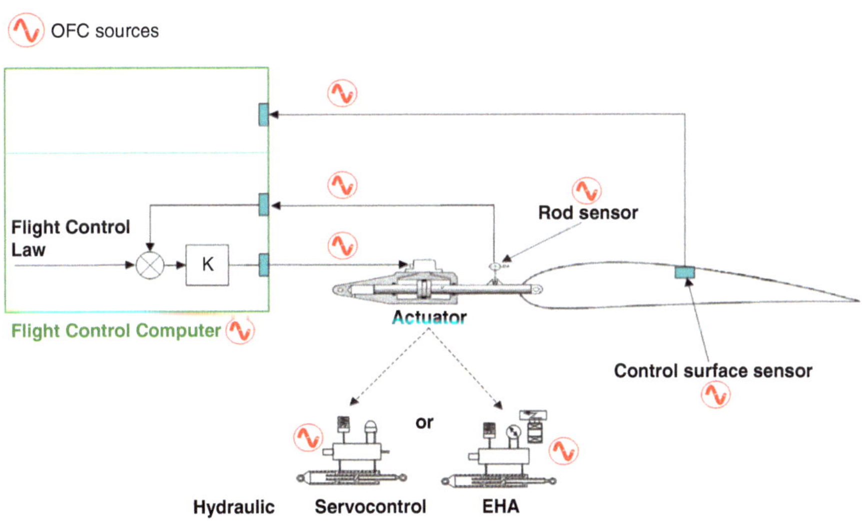

Fig. 3.3 OFC source location within the actuator control loop

detection is expressed in period numbers, which means that the time allowed for detection is not the same depending on the failure frequency.

3.2.2 State-of-Practice: In-Service A380 Aircraft Example

The methodology used to solve OFC problem by Airbus in A380 family is described in [2]. The technique corresponds to an open-loop model-based approach. To the best of our knowledge, it is probably the first time that the concept of analytical redundancy is implemented on a large scale on board for a family of civil aircraft. The fault indicating signal is the difference between the measured control surface position and the estimated position. A nonlinear hydraulic actuator model is used to estimate the position.

3.2.2.1 Nonlinear Hydraulic Actuator Model

The nonlinear model is based on the physical behavior of the hydraulic actuator. The objective is to express the actuator rod speed as a function of the hydraulic pressure delivered to the actuator and the forces applying on the control surface and reacted by the actuator. The two main contributors are aerodynamics forces and the servo-control load in damping mode of the passive actuator in the case of two actuators per control surface.

The actuator rod speed can be expressed by the following basic physical model [2]:

$$\dot{y}(t) = V_0(t).\sqrt{\frac{\Delta P(t) - \frac{F_{\text{aero}}(t)+F_{\text{damping}}(t)}{S}}{\Delta P_{\text{ref}}}} \tag{3.1}$$

where y is the rod position, $\Delta P(t)$ is the hydraulic pressure delivered to the actuator, and $F_{\text{aero}}(t)$ represents the aerodynamic forces applied on the control surface. F_{aero} is a function of the dynamic pressure, the angle of attack, the Mach number, the roll velocity, and the rod position of the actuator. The corresponding model will not be detailed here as it is not of primary interest in this book. $F_{\text{damping}}(t)$ represents the servo-control load of the adjacent actuator in damping mode:

$$F_{\text{damping}}(t) = K_a(t)\dot{y}(t)^2, \tag{3.2}$$

where $K_a(t)$ is the actuator damping coefficient. S is the actuator piston surface area. ΔP_{ref} is the differential pressure corresponding to the maximum rod speed. This speed is reached when the actuator servo-valve is fully opened, i.e., when $\Delta P(t) = \Delta P_{\text{ref}}$. $V_0(t)$ is the rod speed computed by the flight control computer. It corresponds to the maximal speed of the actuator alone with no load

$$V_0(t) = K_{\text{ci}}K(u(t) - y(t)), \tag{3.3}$$

where $u(t)$ is the actuator command signal. K is the servo-control gain, and an estimated current $i(t) = K(u(t) - y(t))$ expressed in milliamp (servo-loop current derived from the flight control law order) is converted in rod speed $V_0(t)$ by a servo-valve model which can be just a simple or double slope gain K_{ci}.

From (3.1), (3.2), and (3.3), the actuator speed (for the hydraulic servo-control) can also be expressed by the following deterministic state space model:

$$\dot{x}(t) = K_{ci}K(u(t) - x(t))\sqrt{\frac{\Delta P(t) - \frac{F_{aero}(t)}{S}}{\Delta P_{ref} + \frac{K_a(K_{ci}K(u(t) - x(t)))^2}{S}}}, \tag{3.4a}$$

$$y(t) = x(t). \tag{3.4b}$$

Note that only the positive sign of the square root is used because of physical correspondence. Moreover, note also that the aerodynamic forces applied on the control surface depend on the sign of the command current, to take into account the influence of the air movement. This feature introduces a sharp nonlinearity that should be taken into account. However, as it will be shown in the following, FDD tuning parameters can be used to compensate the effect the aerodynamic forces could have on the FDD filter output. So, in the following we consider the model 3.4 for further developments. Different saturations (actuator limit positions, maximum orders...) can also be taken into account.

3.2.2.2 Fault Detection

The overall method consists in two steps: residual generation and evaluation [2] (see Fig. 3.4).

The residual is generated by comparing the real position y of the control surface with an estimated position produced by the nonlinear model (3.1) (see Fig. 3.5). In order to reduce the computational burden, some simplifications are performed on the dynamic behavior of hydraulic actuators: K_a and ΔP are assumed to keep constant values, and F_{aero} is considered to be equal to zero [2]. These parameters depend on varying operational conditions such as fluid temperature or number of actuators used simultaneously on a given hydraulic circuit, which make the estimation process difficult. On the A380 onboarded solution, these three parameters are assumed to be constant. This simplification is justified on this aircraft because the achieved detection performances and robustness are compliant with the structural design objectives.

The residual signal is then decomposed in several spectral sub-bands. The OFC detection is performed in each sub-band by counting oscillations of the filtered residual (counting successive and alternate crossings of a given threshold η; see Fig. 3.6). The detectable failure amplitude depends on the model quality. Liquid OFC can be detected by counting around zero alternate and successive crossings

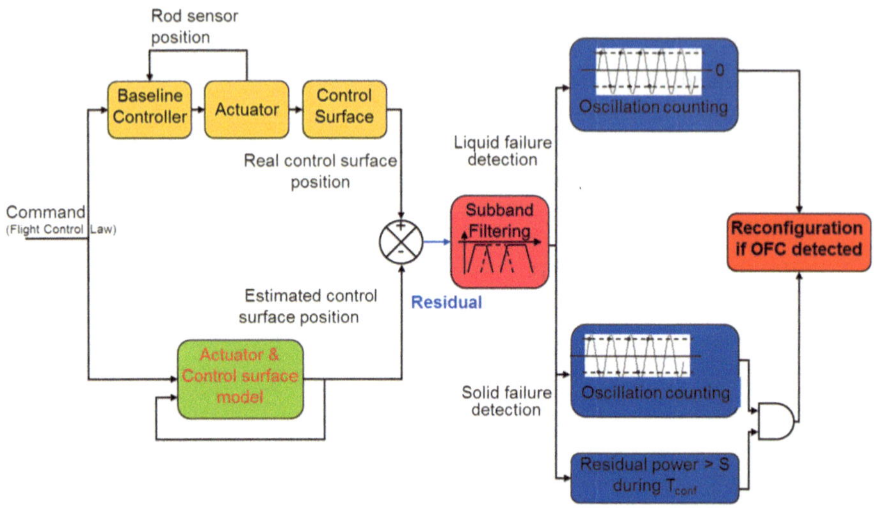

Fig. 3.4 OFC detection in A380 airplane

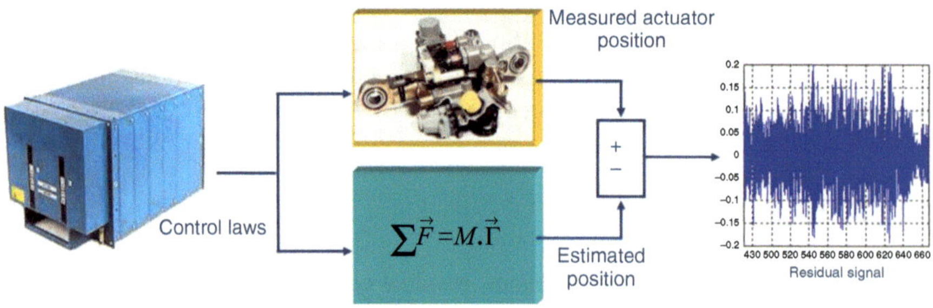

Fig. 3.5 OFC residual generation in A380 airplane

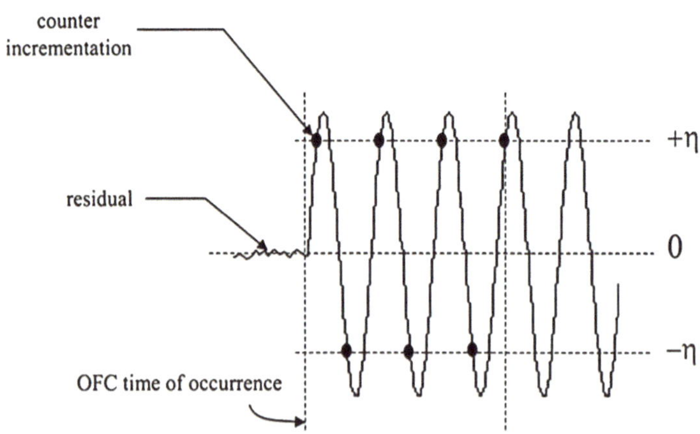

Fig. 3.6 Residual crossing counting illustration

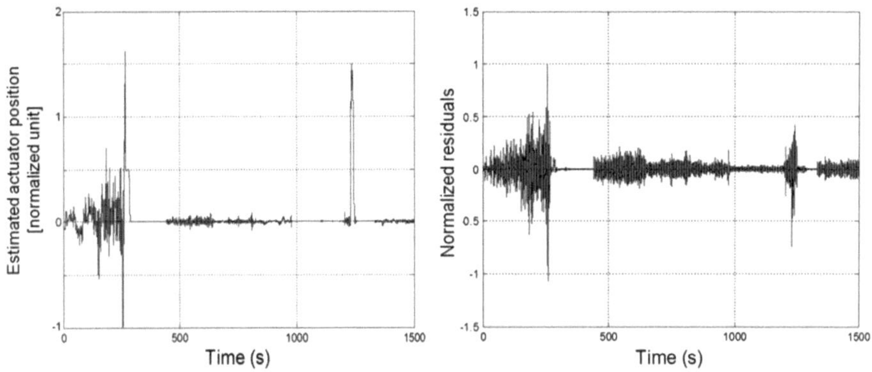

Fig. 3.7 Real surface position and corresponding residual

of the threshold and solid OFC by counting around the opposite of the estimated position. For further details, the interested reader can refer to [2].

The above detection technique has been first validated using real flight data and during severe simulation campaigns on industrial flight simulators. The robustness has been validated on the same test facilities and additionally during several hundred hours of flight test on four A380 aircrafts (Toulouse, France). The technique is currently used onboard in the A380 electrical flight control system, providing a complete OFC coverage without false alarms. At the end of January 2013, 97 A380 aircraft were in service, corresponding to a total of 856,000 flight hours completed during about 104,000 cycles (a cycle is a complete flight between take-off and landing). No false alarms have been noticed, which definitely proved the robustness of the method considering the operating time.

3.2.2.3 A Flight Test Example

An example of real A380 left inboard elevator position and the corresponding residual are depicted in Fig. 3.7.

In this example no OFC signal appears during the flight, and the residual has been generated as described in the previous section.

3.2.3 *Motivations for an Advanced Model-Based Approach*

The monitoring technique for OFC detection described in the previous section is industrially well mastered and well characterized (high level of robustness and good performance). It follows that any modification to this technique should provide, first of all, a viable technological solution ensuring better performance while guaranteeing the same level of robustness. A potential detection method should comply with stringent operational conditions in terms of trade-offs for worst-case detection performance, fault coverage, robustness in harsh environment, computational burden (memory storage, CPU load), and design complexity. It must also offer the possibility of reuse (or building around it), with adequate design and tuning engineering tools. Also, use of approaches with restricted high-level tuning parameters is very important to reduce the test phase needed for aircraft algorithm certification procedure. The final goal is to anticipate the more and more stringent requirements which would come in force for future programs.

Although some published papers exist on estimation of the unknown values of an oscillatory signal (see, for instance, [4–6]); there are few published papers which deal directly with the problem of OFCs. Probably one of the first published works on OFC is [7] where a set of methods called OFIS (Oscillatory Failure Identification System) was presented. These methods correspond to different fault situations and are based on a combination of linear methods and signal processing. It is clear that successful OFC detection cannot be guaranteed when some key assumptions are violated (e.g., linearity or signal-to-noise ratio). Moreover, in practice, it is desirable to have a single method, which is able to cover all fault situations. The motivation behind the results presented in Sects. 3.4 and 3.5 is to develop robust model-based monitoring strategies to detect such failures with small amplitude at an early stage. Before doing that, we start by presenting the tools which will be used for validation of the proposed techniques.

3.3 Verification and Validation Tools

The techniques developed in this chapter are tested and validated on three different V&V means:

- A Matlab/Simulink aircraft model, termed Airbus Aircraft Benchmark (AAB)
- An industrial Airbus actuator bench, also called Airbus Test Facilities (ATF)
- An Airbus flight simulator (cf. Fig. 3.9, picture on the right-hand side) that can be coupled to the ATF

The AAB is a highly representative benchmark developed by Airbus during the European COFCLUO[1] (2007–2010) and ADDSAFE[2] (2009–2012) projects; its structure is given in Fig. 3.8 (see [8–10] for more details). The benchmark is developed within Matlab/Simulink environment.

The AAB is used as the first step for testing and tuning the FDD techniques. The next step of validation is performed on a Functional Engineering Simulator (FES), also developed during ADDSAFE project. The FES provides a faithful simulation environment for the selected fault scenarios and to support the development and

[1]Clearance of flight control laws using optimization.

[2]Advanced fault diagnosis for Sustainable Flight Guidance and Control (http://addsafe.deimos-space.com).

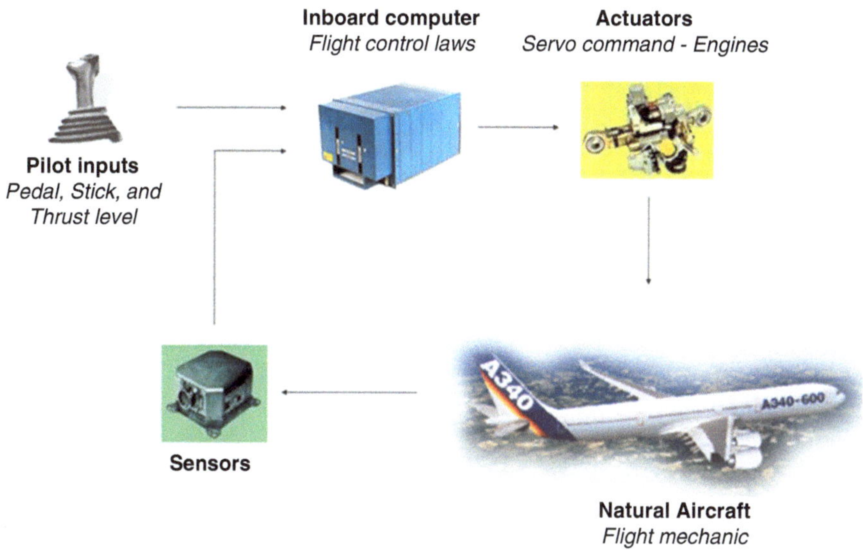

Fig. 3.8 Airbus Aircraft Benchmark

Fig. 3.9 Airbus actuator test bench and A380 flight simulator

benchmarking of the FDD designs [9, 10]. So, the FES can be viewed as an additional layer around the AAB that can ease the exhaustiveness of the V&V activities. In particular, the FES can perform intensive Monte Carlo campaigns for assessing the robustness and performances of FDD designs.

In some cases, the validation is done using real recorded flight data set. The industrial evaluation is done using hardware-in-the-loop simulations with a flight actuator test bench as the one presented in Fig. 3.9. From now on, through this book, we will call it Airbus Test Facilities (ATF). The test bench is built around a real control surface actuator with simulated command inputs, aerodynamic forces, and hydraulic pressures. This bench offers also the possibility to validate the designed system in degraded configurations, as in the case of low hydraulic pressure and high loads on the control surface.

The industrial validation campaign consist mainly of the assessment of the robustness (i.e., the lack of false alarms) and of the detection performance (i.e., the lack of missed detections and satisfactory detection time). The robustness can

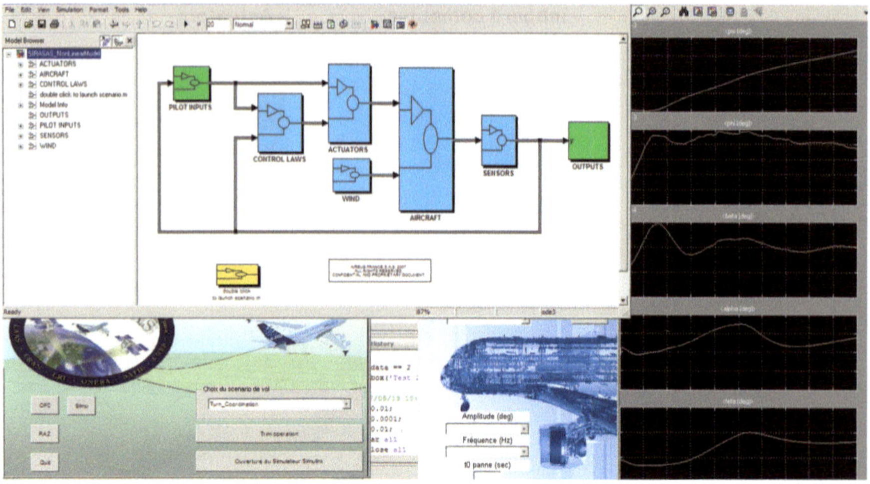

Fig. 3.10 Airbus Aircraft Benchmark

be assessed during pure lateral maneuvers, pure longitudinal, and during mixed maneuvers (combining lateral and longitudinal movement in the same maneuvers). Both smooth and dynamic maneuvers can be performed, as, for example, autopilot maneuvers, fight control checks, take-off, and landing.

3.3.1 Airbus Aircraft Benchmark (AAB)

The benchmark is based on six degrees of freedom dynamic aircraft model, defined in Matlab/Simulink. It includes aerodynamic, engine, atmospheric, and gravity models. In addition, actuator and sensor characteristics are taken into account, together with models for external disturbances. The flight control laws have longitudinal and lateral components, each of which contains inner and outer loops, as with a conventional autopilot. Once a trim condition is established within the simulation environment, a user interface allows the user to simulate a number of flight scenarios that can be executed in healthy and faulty situations (Fig. 3.10).

3.3.2 Functional Engineering Simulator (FES)

The Functional Engineering Simulator (FES), developed by Deimos Space[3] during ADDSAFE project, is a non-real-time simulator based on Simulink, Matlab, Spain and XML that includes AAB as well as robustness and performance analysis tools

[3]http://www.deimos-space.com/en/

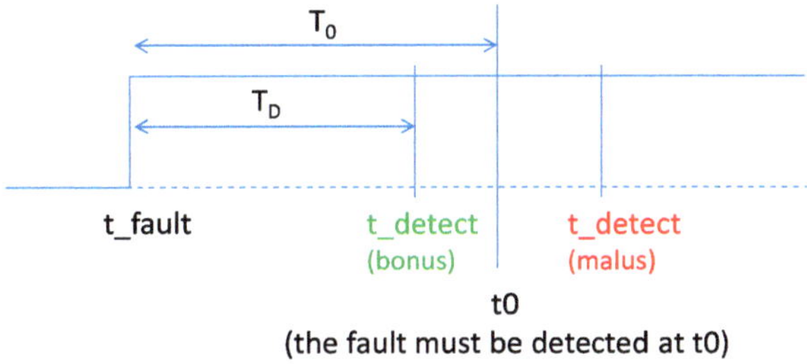

Fig. 3.11 Definition of fault and detection time parameters

for all the fault scenarios defined in the project [10, 11]. FES is a term used in Space to describe a software simulator describing at a functional level the components of a system (including its operating environment). FES are used in support of the specification, design, verification and operations of space systems, and can be used across the spacecraft development life-cycle, including activities such as system design validation, software verification & validation, spacecraft unit and sub-system test activities. Once a simulation has been run, the raw simulation outputs can be post-processed to obtain new variables for the analysis of the system. Figures-of-Merit (FOM) are produced as scalar quantities that characterize the performance of the FDD system.

3.3.3 *Industrial Assessment Criteria*

The purpose of this section is to list the quantitative and qualitative criteria used to evaluate FDD designs. The quantitative component is given by metrics and a cost function which is automatically calculated by the FES for a given FDD design based on a Monte Carlo campaign. The qualitative evaluation is used to assess the designs' practical implementation and relevance for industrial use.

3.3.3.1 Quantitative Assessment

In order to define an evaluation matrix, a set of definitions are first presented based on Fig. 3.11.

In Fig. 3.11 the definition for each "time" parameter is as follows:

- t_fault is the time instant at which a fault is activated.
- t_detect is the time instant at which a fault is declared as detected by the FDD system.
- $t0$ is the specified time instant at which a fault must have been declared by the FDD system.
- T_D is the detection time that is the time period between activation of a fault and its detection.
- T_0 is the maximum allowed detection time.

With these definitions, the evaluation metrics can be defined. The Detection Time Performance (DTP) can be defined as

$$\mathrm{DTP} = \frac{T_D}{T_0} = \frac{\text{t\_detect} - \text{t\_fault}}{t_0 - \text{t\_fault}}. \tag{3.5}$$

Statistics of this metric, such as average, minimum, maximum, and variance values are calculated. The false alarm (FA) rate metric is computed taking into account the total number of cases yielding a false alarm n_{FA} out of the total number of Monte Carlo runs n_{MC}

$$\mathrm{FA}_{\%} = \frac{n_{\mathrm{FA}}}{n_{\mathrm{MC}}} 100. \tag{3.6}$$

In the same manner, the missed detection (MD) rate index is computed as the percentage ratio of MD cases n_{MD} versus n_{MC}:

$$\mathrm{MD}_{\%} = \frac{n_{\mathrm{MD}}}{n_{\mathrm{MC}}} 100. \tag{3.7}$$

An executive time (ET) metric is defined to represent the computational burden of the proposed designs. This metric is more oriented toward the FDD design viability (e.g., applicability) within a real-time environment. This metric is very useful for estimating the required percentage of the FCC CPU that is demanded for an implementation. These four metrics are the main criteria used for assessing the robustness and performances of the proposed designs.

Note that only the maximum DTP is retained as a significant statistic for the cost function computation. Indeed, in view of a possible use in practical aerospace applications, the worst case must be taken always into account rather than an average behavior.

3.3.3.2 Qualitative Assessment

A pure quantitative assessment (3.5), (3.6), and (3.7) is obviously not sufficient for evaluation of the industrial relevance of the proposed FDD designs. Some additional qualitative metrics can be used, as, for example, the number of input parameters of the FDD designs, with or without a physical meaning. This metric is of primary interest as it impacts the V&V workload and consequently the system development duration. In view of adapting the proposed design to another control surface and/or on a different aircraft, it is more interesting to use input parameters with a physical meaning. It is also industrially relevant to know if a clear procedure for step-by-step tuning of the FDD design can be specified describing, for example, how many steps are required for a high-level tuning. Among the number of input parameters, the number of parameters to tune is also an important criterion. In fact, many available design methods are not really associated with clear "tuning" guidelines: a simple and rudimentary well-mastered method may work quite better than a complex design method, if the end user cannot tune it properly. Note that the workload of V&V activities is also impacted as the effect of a bad tuning must be monitored.

3.4 Nonlinear Observer Design

In the previous section, the A380 OFC detection method has been briefly discussed. In this section, a nonlinear observer-based OFC detection algorithm is proposed. The purpose, as already mentioned, is the detection of small OFCs which may go undetected with the method described in Sect. 3.2, while ensuring a good robustness level. The material of this section is partly underpinned by the published paper [3].

3.4.1 *OFC Detectability*

To start, let us look at the problem of OFC detection from a detectability point of view. Oscillatory failures are modeled by the effect they could have on the command signal of actuators. In the following, the fault-free command signal is represented by $u_o(t)$. The effect of OFC is modeled as an additive term to the command signal; this term is represented by $f(t)$ whose specific value depends on the kind of fault (liquid or solid) to be modeled, i.e., $f(t) = A_l \sin(\omega_l t)$ if the fault is liquid and $f(t) = A_s \sin(\omega_s t) - u_o(t)$ if the fault is solid. If no fault is present, then $f(t) = 0$ and $u(t) = u_o(t)$. OFC detectability could be analyzed using sensitivity function [12]. Consider the following definition of detectability:

Definition 3.1 Consider an OFC modeled by setting $u(t) = u_o(t) + f(t)$. An OFC is said to be structurally detectable if for some $u_o(t)$

$$\frac{\partial y}{\partial f} \neq 0.$$

The OFC detectability condition can now be formulated as follows.

Proposition 3.1 *The OFC associated with the system (3.4) is detectable in the sense of the above detectability definition.*

Proof Consider the system with faults

$$\begin{cases} \dot{x}(t) = F(x(t), u_0(t) + f(t)) \\ y(t) = x(t) \end{cases}.$$

Let us define

$$\left.\frac{\partial y}{\partial f}\right|_{f=0} = \left.\frac{\partial x}{\partial f}\right|_{f=0} = \sigma.$$

The variation of (3.4) with respect to f can be also calculated as

$$\frac{\partial}{\partial f}\dot{x} = \frac{\partial}{\partial f}F(x,u)$$

or

$$\frac{\mathrm{d}}{\mathrm{d}t}\frac{\partial x}{\partial f} = \frac{\partial F}{\partial x}\frac{\partial x}{\partial f} + \frac{\partial F}{\partial u}\frac{\partial u}{\partial f}.$$

Since

$$\frac{\partial u}{\partial f} = 1,$$

we get

$$\dot{\sigma} = \left.\frac{\partial F}{\partial f}\right|_{x(t)}\sigma + \left.\frac{\partial F}{\partial f}\right|_{x(t)}.$$

In order to show that $\sigma \neq 0$ for $t > t_0$, first note that $\sigma(t_0) = 0$ and the above system is scalar linear time varying. Note that initial conditions of $\sigma(t_0)$ are zero, so the state $\sigma(t)$ could be different from zero only if a forcing function is acting on the differential equation. The forcing function is represented by $\partial F/\partial f$. It is easy to see that the differential equation for $\sigma(t)$ is controllable (considering $\partial F/\partial x$ as system matrix and 1 as the coefficient to the input). So, $\sigma(t) \neq 0$ if the input $\partial F/\partial f = \partial F/\partial u$ is also not equal to zero

$$\frac{\partial F}{\partial u} = K_{\mathrm{ci}}K\left(\frac{\Delta P}{\Delta P_{\mathrm{ref}} + \frac{K_a K_{\mathrm{ci}}{}^2 K^2(u-z)^2}{S}}\right)^{1/2} + \frac{K_{\mathrm{ci}}K(u-z)}{2S}(-1).$$

$$\left(\frac{\Delta P}{\Delta P_{\mathrm{ref}} + \frac{K_a K_{\mathrm{ci}}{}^2 K^2(u-z)^2}{S}}\right)^{-1/2} \cdot \left(\frac{\Delta P\left(2\frac{K_a}{S}K_{\mathrm{ci}}{}^2 K^2(u-z)\right)}{\left(\Delta P_{\mathrm{ref}} + \frac{K_a K_{\mathrm{ci}}{}^2 K^2(u-z)^2}{S}\right)^2}\right)$$

and after some manipulations

$$\frac{\partial F}{\partial u} = K_{\mathrm{ci}}K\left[1 - \frac{K_a K_{\mathrm{ci}}{}^2 K^2(u-z)^2}{\Delta P_{\mathrm{ref}} + \frac{K_a K_{\mathrm{ci}}{}^2 K^2(u-z)^2}{S}}\right]\left(\frac{\Delta P}{\Delta P_{\mathrm{ref}} + \frac{K_a K_{\mathrm{ci}}{}^2 K^2(u-z)^2}{S}}\right)^{1/2}.$$

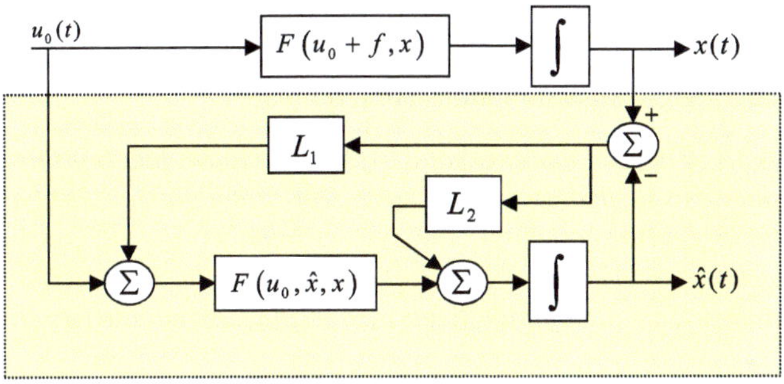

Fig. 3.12 Bloc diagram of the nonlinear residual generator

Defining a positive constant $\mu_3 > 0$, the following inequalities could be established (see proof of Lemma 3.2)

$$0 < \mu_3 < \left(1 - \frac{K_a {K_{ci}}^2 K^2 (u-x)^2}{S\Delta P_{ref} + K_a {K_{ci}}^2 K^2 (u-x)^2}\right)\left(\frac{\Delta P}{\Delta P_{ref} + \frac{K_a {K_{ci}}^2 K^2 (u-x)^2}{S}}\right)^{1/2} = \frac{\partial F}{\partial u}$$

using this inequality $\partial F/\partial u < -K_{ci} K \mu_3$. So that if $\sigma \neq 0$ any $f \neq 0$ will be manifested on the output. This completes the proof.

With the above analysis in mind, in the following section the residual generator based on a nonlinear observer approach is presented.

3.4.2 *Proposed Detection Algorithm*

Consider the system to be supervised given by (3.4). The proposed residual generator for this system is given by

$$\dot{\hat{x}} = F(\hat{x},\ u_o + L_1\tilde{x}) + L_2\tilde{x}, \tag{3.8}$$

$$r = y - \hat{x} \tag{3.9}$$

where $r = \tilde{x} = y - \hat{x}$ with $L_1 \in R$ and $L_2 \in R$. The observer gains $L_i,\ i = 1, 2$ are defined to be linear and constant. This kind of observer with $L_1 = 0$ is frequently called Thau-like observer, because Thau proposed in [13] to use linear constant observer gain. However, the proposed residual does not behave exactly like a conventional Thau-like observer because of the internal loop given by the gain L_1. A schematic block diagram of the proposed observer-based residual is given in Fig. 3.12.

3.4.2.1 Stability Analysis

To proceed, the following assumptions are required:

Assumption 3.1 *The variables $u(t)$ and $x(t)$ are bounded functions for all $t \in R_+$. This assumption is satisfied because of some software logic and dedicated monitoring in the flight control computer; the system input $u(t)$ is always bounded. For the same reasons and because of physical restrictions, the actuator position $x(t)$ is actually bounded. This assumption leads also to the fact that the function $F(x,u)$ is bounded.*

Assumption 3.2 *Without loss of generality, the term $\partial u_o/\partial x$ is assumed to be always negative. This assumption corresponds to a negative feedback, and it is actually natural.*

Some mathematical tools necessary to prove the main result are given below.

Theorem 3.1 (*Mean Value Theorem*). *Assume that F is a continuous function everywhere on the closed interval $[a,b]$ and has a derivative in each point of the open interval (a,b). Then there is at least one interior point c of (a,b) for which*

$$F(b) - F(a) = F'(c)(b-a),$$

where

$$F'(c) = \left.\frac{\partial F}{\partial x}\right|_{x=c}.$$

Proof See, for example, [14].

Lemma 3.1 *Grönwall-Bellman Inequality. Let I denote an interval of the real line of the form $[a,b)$ with $a < b$. Let β and u be real-valued continuous functions defined on I. If u is differentiable in the interior of I° (I° is the interval without the end points a and b) and the differential inequality is satisfied*

$$\dot{u}(t) \leq \beta(t)u(t), \quad t \in I^{\circ},$$

then u is bounded by the solution of the corresponding differential equation

$$u(t) \leq u(a) \exp\left(\int_a^t \beta(s)\mathrm{d}s\right)$$

for all $t \in I$.

Proof See, for example, [15].

Lemma 3.2 *Assume* $\Delta P_{\text{ref}} > 0$; $\Delta P > 0$; $K_{\text{ci}} > 0$; $K > 0$; $K_a > 0$; $S > 0$, *and* $F_{\text{aero}} = 0$. *Consider the Lipschitz function* $F(x, u)$. *Then*

$$\left.\frac{\partial F}{\partial x}\right|_{x=z} < -K_{\text{ci}} K \mu$$

with $\mu > 0$.

Proof The partial derivative of $F(x(t), u(t))$ with respect to $x(t)$ results in

$$\left.\frac{\partial F}{\partial x}\right|_{x=z} = \frac{\partial}{\partial u}\left[K_{\text{ci}} K(u-x)\left(\frac{\Delta P}{\Delta P_{\text{ref}} + \frac{K_a K_{\text{ci}}^2 K^2 (u-x)^2}{S}}\right)^{1/2}\right]\frac{\partial u}{\partial x}$$
$$+\frac{\partial}{\partial x}\left[K_{\text{ci}} K(u-x)\left(\frac{\Delta P}{\Delta P_{\text{ref}} + \frac{K_a K_{\text{ci}}^2 K^2 (u-x)^2}{S}}\right)^{1/2}\right]_{x=z}$$

$$\left.\frac{\partial F}{\partial x}\right|_{x=z} = K_{\text{ci}} K(u-x)\left(\frac{\Delta P}{\Delta P_{\text{ref}} + \frac{K_a K_{\text{ci}}^2 K^2 (u-z)^2}{S}}\right)^{1/2}\frac{\partial u}{\partial x}$$
$$+\frac{K_{\text{ci}} K(u-z)}{2S}\frac{\partial u}{\partial x}(-1)\left(\frac{\Delta P}{\Delta P_{\text{ref}} + \frac{K_a K_{\text{ci}}^2 K^2 (u-z)^2}{S}}\right)^{-1/2}$$
$$\left(\frac{\Delta P\left(2\frac{K_a}{S} K_{\text{ci}}^2 K^2 (u-z)\right)}{\left(\Delta P_{\text{ref}} + \frac{K_a K_{\text{ci}}^2 K^2 (u-z)^2}{S}\right)^2}\right) + (-1) K_{\text{ci}} K\left(\frac{\Delta P}{\Delta P_{\text{ref}} + \frac{K_a K_{\text{ci}}^2 K^2 (u-z)^2}{S}}\right)^{1/2}$$
$$+\frac{K_{\text{ci}} K(u-z)}{2S}\left(\frac{\Delta P}{\Delta P_{\text{ref}} + \frac{K_a K_{\text{ci}}^2 K^2 (u-z)^2}{S}}\right)^{-1/2}\left(\frac{\Delta P\left(2\frac{K_a}{S} K_{\text{ci}}^2 K^2 (u-z)\right)}{\left(\Delta P_{\text{ref}} + \frac{K_a K_{\text{ci}}^2 K^2 (u-z)^2}{S}\right)^2}\right).$$

Defining a positive variable $\mu_1(t) > 0$ where $\partial u_o(t)/\partial x(t) = -\mu_1(t)$ and after some manipulations we get

$$\left.\frac{\partial F}{\partial x}\right|_{x=z} = -\left[K_{\text{ci}} K - \frac{K_a K_{\text{ci}}^3 K^3 (u-z)^2}{\Delta P_{\text{ref}} + \frac{K_a K_{\text{ci}}^2 K^2 (u-z)^2}{S}}\right]\left(\frac{\Delta P}{\Delta P_{\text{ref}} + \frac{K_a K_{\text{ci}}^2 K^2 (u-z)^2}{S}}\right)^{1/2}\mu_1(t)$$
$$-\left[K_{\text{ci}} K - \frac{K_a K_{\text{ci}}^3 K^3 (u-z)^2}{\Delta P_{\text{ref}} + \frac{K_a K_{\text{ci}}^2 K^2 (u-z)^2}{S}}\right]\left(\frac{\Delta P}{\Delta P_{\text{ref}} + \frac{K_a K_{\text{ci}}^2 K^2 (u-z)^2}{S}}\right)^{1/2}.$$

Using this relation and after some further manipulations, the right hand of the above equation could be rewritten as

$$\left.\frac{\partial F}{\partial x}\right|_{x=z} = -K_{\text{ci}}K(\mu_1(t)+1)\left(\frac{\Delta P}{\Delta P_{\text{ref}} + \frac{K_a K_{\text{ci}}{}^2 K^2 (u-z)^2}{S}}\right)^{1/2}$$
$$\times \left(1 - \frac{K_a K_{\text{ci}}{}^2 K^2 (u-z)^2}{S\Delta P_{\text{ref}} + K_a K_{\text{ci}}{}^2 K^2 (u-z)^2}\right).$$

Note that because of the technical assumptions given in the lemma, the following bounds could be established:

$$0 < \left(\frac{\Delta P}{\Delta P_{\text{ref}} + \frac{K_a K_{\text{ci}}{}^2 K^2 (u-z)^2}{S}}\right)^{1/2} < 1$$
$$1 - \frac{K_a K_{\text{ci}}{}^2 K^2 (u-z)^2}{S\Delta P_{\text{ref}} + K_a K_{\text{ci}}{}^2 K^2 (u-z)^2} > 0$$

$\forall u(t), z(t)$. Consequently there exists a constant $\mu > 0$ such that

$$0 < \mu \le \left(1 - \frac{K_a K_{\text{ci}}{}^2 K^2 (u-z)^2}{S\Delta P_{\text{ref}} + K_a K_{\text{ci}}{}^2 K^2 (u-z)^2}\right)\left(\frac{\Delta P}{\Delta P_{\text{ref}} + \frac{K_a K_{\text{ci}}{}^2 K^2 (u-z)^2}{S}}\right)^{1/2}$$
$$(\mu_1(t)+1).$$

Substituting the time-varying terms of the right by a lower bound, the claimed inequality is obtained and this completes the proof.

Note that, to get the result of Lemma 3.2, no explicit knowledge of the parameters values is needed.

The main result is now given.

Theorem 3.2 *The system* (3.8), (3.9) *is a residual generator for the actuator* (3.4).

Proof In order to show that the proposed equation system (3.8), (3.9) is a residual generator, it should be proved that in the fault-free case, the residual converges to zero, and when OFCs are present, the residual is not zero. Consider first the fault-free case, i.e., $u(t) = u_o(t)$. The residual is defined in a conventional way as the output estimation error

$$\tilde{y}(t) = y(t) - \hat{y}(t)$$

but because $y(t) = x(t)$

$$\tilde{x}(t) = x(t) - \hat{x}(t)$$

and so

$$\dot{\tilde{x}}(t) = \dot{x}(t) - \dot{\hat{x}}(t).$$

Substituting the corresponding equations of the system (3.4) and the residual generators (3.8) in the last equation, the dynamics error equation results in

$$\dot{\tilde{x}}(t) = F(x(t),\ u(t)) - F(\hat{x}(t),\ \hat{u}(t) + L_1\tilde{x}(t)) - L_2(x(t) - \hat{x}(t)). \quad (3.10)$$

Now, to obtain a more convenient form for the error, the mean value theorem is applied to (3.10), i.e., to the first two terms of the right side. Observing that because of the special form of $F(x(t), u(t))$ given in (3.4a), the second term could be written as $F(\hat{x} + L_1\tilde{x})$ and for the fault-free case $u(t) = u_o(t)$, so the first two terms of the right side of (3.4) result

$$F(x,\ \hat{u}) - F(\hat{x} + L_1\tilde{x},\ \hat{u}) = \left.\frac{\partial F(x,\ u)}{\partial x}\right|_{z(t)} (1 + L_1)\,\tilde{x}.$$

The resulting error dynamics becomes

$$\dot{\tilde{x}}(t) = -\left(L_2 - \left.\frac{\partial F(x,\ \hat{u})}{\partial x}\right|_{z(t)} (1 + L_1)\right)\tilde{x}(t),$$

where $z(t)$ is unknown. Note that the unknown $z(t)$ should be considered time variant because $x(t)$ is also time variant; however, it should be inside the open interval $(x(t), \hat{x}(t))$. The dynamics of the error results in a linear time-variant unforced system. Using the result of Lemma 3.2, the dynamics error equation is given by

$$\dot{\tilde{x}}(t) < -(L_2 + K_{\mathrm{ci}}K\mu(1 + L_1))\,\tilde{x}(t)$$

and consequently, using Lemma 3.1, the estimation error satisfies

$$\dot{\tilde{x}}(t) < e^{-(L_2 + K_{\mathrm{ci}}K\mu(1+L_1))}\tilde{x}(0),$$

i.e., the estimation error converges to zero if no OFC is present.

Now, consider the presence of OFCs in the system. An OFC is modeled as an additional term to the input, and it is represented by f. The dynamics of the estimation error are given by

$$\dot{\tilde{x}}(t) = F(x(t),\ \hat{u}(t) + f) - F(\hat{x}(t),\ \hat{u}(t) + L_1\tilde{x}(t)) - L_2(x(t) - \hat{x}(t)). \quad (3.11)$$

Considering the addition and rest of the term $F(x(t), u_0(t)$ in (3.11) gives

$$\begin{aligned}\dot{\tilde{x}}(t) = {} & F(x(t), \hat{u}(t) + f) - F(x(t), \hat{u}(t)) + F(x(t), \hat{u}(t)) \\ & - F(\hat{x}(t), \hat{u}(t) + L_1\tilde{x}) - L_2(x(t) - \hat{x}(t)). \end{aligned} \quad (3.12)$$

Again, applying the mean value theorem to the first two terms of (3.12), observing that only a variable associated to $u(t)$ is changing, we obtain

$$F(x(t), u_0(t) + f) - F(x(t), u_0(t)) = \left.\frac{\partial F}{\partial f}\right|_{\varsigma} f.$$

Defining $g(x(t), u_o(t), \varsigma(t)) \triangleq \left.\frac{\partial F}{\partial f}\right|_{\varsigma}$ and applying the mean value theorem to the following two terms, using the results of Lemma 3.2 (note that the rest of the terms are just similar to the ones in (3.10)), the dynamics of the estimation error $\tilde{x}(t)$ is given by

$$\dot{\tilde{x}}(t) = -\left(L_1 - \left.\frac{\partial F(x, \hat{u})}{\partial x}\right|_{z(t)} (1 - L_1)\right) \tilde{x}(t) + g(x(t), u_0(t), \varsigma(t)) f.$$

So, under the presence of OFC (represented by f) the estimation error is a forced, time-variant differential equation, i.e., the estimation error depends on the forcing term (the OFC). Now, if $f \neq 0$, the estimation error is not zero and the proposed observer could be used as a residual generator to detect OFCs. This completes the proof.

As it can be seen from the proof of Theorem 3.2, the dynamics of the residual, when no OFC is present, is a nonlinear homogeneous equation, and the gains of the residual (L_1 and L_2) can be selected in order to have residual zero (ideal case) under the assumption $F_{\text{aero}} = 0$ and no uncertainties.

Remark 3.1 Consider the residual generator (3.8). If L_1 is chosen as $L_1 = -1$, the error dynamics, i.e., the residual dynamics, becomes linear. The gain L_2 can be designed to adjust convergence time.

Remark 3.2 If $F_{\text{aero}} \neq 0$ and/or uncertainty on ΔP is present, the gains L_1 and L_2 could be selected in order to manage the trade-off between robustness with respect to perturbations/uncertainty and sensitivity to OFC.

In the previous developments, the free parameters to be tuned are L_1 and L_2. Define $L = [L_1, L_2]^{\mathrm{T}}$. The choice of L impacts largely the level of performance that can be obtained. The residual should be robust (not generate false alarms) and, at the same time, sensitive to OFC to be detected. For example, if the residual is very robust, but the effect of faults is attenuated on the residual, it would be useless. On the other hand, from an application and industrial point of view, there is a need to provide high level and formalized tuning tools that can be easily used by non-expert operators.

The robustness can be directly related to the observation error. For the fault detection it is necessary to take into account the observation error in the fault-free case, the sensitivity to faults, the robustness against perturbation, the time of detection, and so on (the evaluation criterions (3.5), (3.6), and (3.7)). Some of these characteristics can be evaluated a posteriori only, after the simulation run, some of them depend on the fault models (like sensitivity to faults, false alarms, and time of detection). Complex nonlinear dependence of fault detection performance on L makes difficult the application of analytical optimization approaches. Here we propose an efficient numerical procedure to solve the problem.

Let k be the sampling time. Assume that we have at hands a data set (e.g., from a flight experiment or recorded from a simulator). Let N be the size of the data set. Denote the sequence of integers $1, \ldots, k$ as $\overline{1,k}$.

The method consists in the following steps:

1. Choose the actuator model and the test data set representing u_{0k} and y_k, $k = \overline{1,N}$, $N > 0$.
2. Choose the models of "the most probable" faults f_k^j, $j = \overline{1,K}$, $K > 0$ with corresponding admissible times of detection τ^j, $j = \overline{1,K}$. For OFC, the most probable faults can be specified in terms of amplitude and frequency.
3. Choose a grid of matrices L^i, $i = \overline{1,M}$, $M > 0$, covering the range of possible values of L.
4. Choose a performance criteria I characterizing fault detection performance.
5. Perform optimization of I over grid L^i, $i = \overline{1,M}$ for chosen γ_k^j, τ^j, $j = \overline{1,K}$ and given $\hat{u}_k$, y_k, $k = \overline{1,N}$.

The model of the actuator is given by (3.4) and the numerical (nominal) values of the parameters are available. The test data set composed by the fault-free control u_{0k}, examples of measurements y_k, and the corresponding time instants t_k are typically available after preliminary experiments. They have to represent the most typical and important operation modes of the actuator. As in the previous sections, the models of faults for OFC case can be chosen as $f_k^j = A_k^j \sin(2\pi\,\omega^j t_k)$, $k = \overline{1,N}$, $j = \overline{1,K}$, where ω^j lies in the range from 1 to 10 Hz (as it was explained above, the frequencies beyond 10 Hz are not considered because they are outside the actuator dynamics bandwidth). The corresponding amplitudes A_k^j depend on the frequencies ω^j and the current amplitude of the fault-free control (for each frequency several amplitudes can be chosen). The amplitudes $A_k^j = 0$ could be chosen for some $k > 0$ to simulate a fault appearance and fading.

The grid of L^i, $i = \overline{1,M}$, $M > 0$ can be chosen using Monte Carlo method, or based on some a priori knowledge on the most representative samples from a given range.

The performance criteria for each L^i, $i = \overline{1,M}$ should include the following representative criteria:

- The output observation error in the fault-free case and the robustness against perturbation, which can be defined as previously for $u_k = u_{ok}$ by the functional

$$J_0^i = N^{-1}\sqrt{\sum_{k=1}^{N} r_k^T r_k}$$

- The sensitivity to faults, which can be characterized by the ratio

$$\frac{J_j^i}{J_0^i}, j = \overline{1, K},$$

where $J_j^i = N^{-1}\sqrt{\sum_{k=1}^{N} r_k^T r_k}$ is computed in the same way, but for $u_k = u_{0k} + f_k^j$ or $u_k = f_k^j$

- The false alarms rate
- The (average) time of detection T_j^i (computed by the fault detection algorithm for each fault f_k^j)

The functional J_0^i has to be minimized, the ratios J_j^i / J_0^i, $j = \overline{1, K}$ have to be maximized (increasing sensitivity), and the detection times T_j^i, $j = \overline{1, K}$ have to satisfy the constraint $T_j^i \leq \tau^j$. Consequently, the fault detection performance can be expressed as follows:

$$I^i = \lambda J_0^i + (1 - \lambda) K^{-1} \sum_{j=1}^{K} J_0^i / J_j^i - \ln(T_j^i \leq \tau^j), i = \overline{1, M},$$

where $0 \leq \lambda \leq 1$ is a constant weight adjusting the influence of the estimation or fault detection terms on the total value of the functional I^i. It is assumed that argument of the logarithm is 1 for $T_j^i \leq \tau^j$ (the condition is true), and the logarithm argument is 0 in the case $T_j^i > \tau^j$, thus $\ln(0) = -\infty$, that penalizes I^i.

Having values I^i of the performance functional on the grid L^i, $i = \overline{1, M}$, its optimization is straightforward:

$$L^* = L^{i*}, i^* = \arg \min_{i=\overline{1,M}} I^i.$$

The nonlinear observer with gain L^* ensures optimal fault detection performance for the chosen grid L^i, the fault models γ_k^j, and the given test data set. The proposed procedure has a simple computer implementation and can be performed off-line for a given actuator.

As an example, using an A380 data set, some results of optimization are illustrated below for some OFC frequencies and amplitudes. The optimal values L^* are in the dark blue region in Fig. 3.13.

3.4.3 *Decision-Making Rule*

The decision function for OFC detection and confirmation is well described in [2]. OFC detection consists in counting successive and alternate crossings of a given

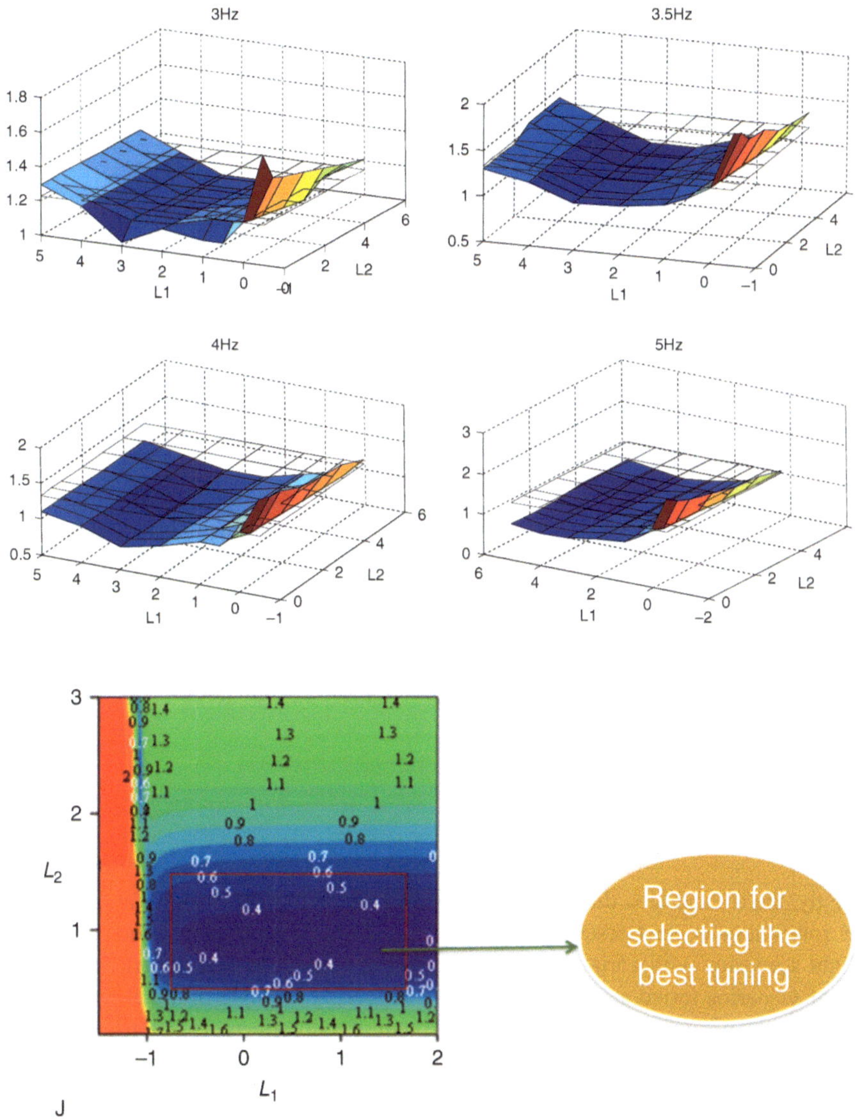

Fig. 3.13 Observer gain optimization

threshold in a sliding time window. Here, the flight control law is considered as fault free. All its oscillations are judged normal and are calculated to compensate for any normal perturbation (e.g., an external disturbance such as turbulence). The hypothesis of a fault-free command is justified because the flight control law is also monitored by dedicated techniques. In case of a solid failure, the OFC substitutes the nominal signal. If the estimated position is null (no control surface deflection), the residual is only composed of the failure, then detection is done by oscillation

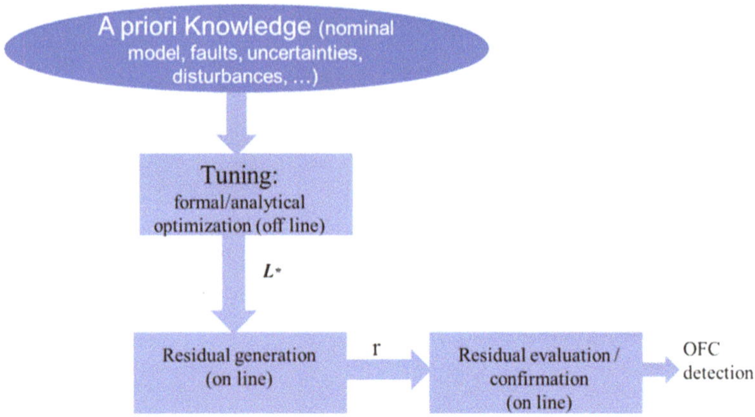

Fig. 3.14 Bloc diagram for OFC detection

counting around zero, like for a liquid failure. But if a control surface deflection is demanded by the flight control law (e.g., during a maneuver or in reaction to the failure), the failure signal is mixed with the opposite of the estimated position, and an oscillation counting around zero would not enable detection. The oscillation counting must be performed around the opposite of the estimated position. Both types of counting, for liquid and solid OFC, are performed in parallel because of the unknown nature of the fault.

3.4.4 Experimental Results

Firstly, the overall bloc diagram is resumed in Fig. 3.14. The performance in terms of robustness is next tested using FES. The tests have been conducted for six flight scenarios with variations in the operating conditions and uncertainties. The flight scenarios tested are a cruise phase, a triggering of angle of attack protection, a so-called nose-up maneuver (abrupt longitudinal movement), a triggering of pitch protection, a coordinated turn, and a "yaw angle mode" which roughly corresponds to an enhanced autopilot hold mode. The simulation campaign for one flight maneuver has been defined with 324 simulation runs that result from the combination of the following parameters:

- Altitude: $h = [8{,}000\ 18{,}000\ 28{,}000\ 38{,}000]$ ft
- Calibrated airspeed: $V_{CAS} = [160\ 220\ 300]$ kts
- Mass: $m = [120{,}000\ 180{,}000\ 233{,}000]$ kg
- X-component of the gravity center: [0.17 0.3 0.41]

For each combination of these flight and aircraft parameters, three additional variations (minimum, nominal, and maximum errors or uncertainties) associated with the aerodynamic coefficients and sensors measurements have also been

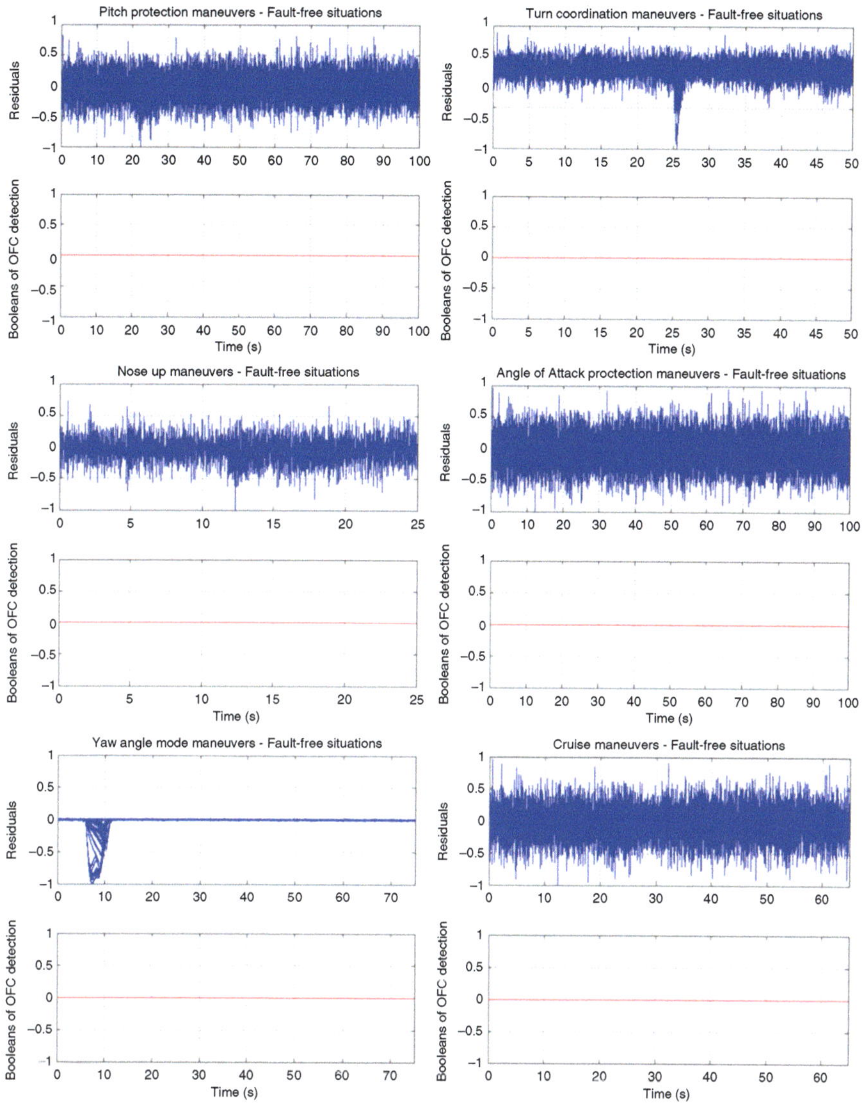

Fig. 3.15 Normalized residuals in fault-free situations with parametric variations: Nonlinear observer

included. Only realistic operating points belonging to the flight envelop are taken into account within the Figures-of-Merit, i.e., only realistic situations are used to assess our FDD system.

In the normal fault-free mode, the FES results show a good robustness against parametric variations, and it gives 0 % of false alarms for the 1,200 realistic simulation runs; the results are depicted in Fig. 3.15. Next, the faulty situations with the parametric variations have been analyzed for the cruise maneuver. The

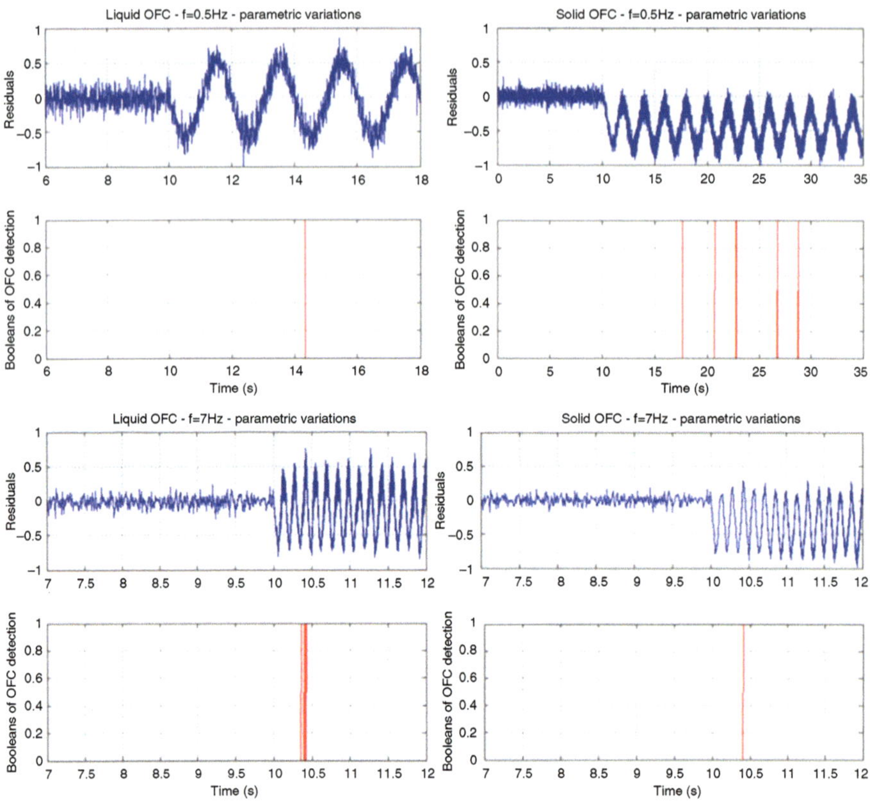

Fig. 3.16 Normalized residuals for liquid (*left*) and solid (*right*) fault

normalized results corresponding to the smallest OFC amplitude and the minimal and maximal OFC frequencies are given in Fig. 3.16. As it can be seen, the residual is small in the fault-free mode, and a significant change in the residual appears in the faulty situation. We get a little bit worse behavior for the solid fault with the frequency 0.5Hz (see Fig. 3.16, the left-bottom plot), which can be explained by a bad estimation of the actuator position under this fault.

The proposed scheme has been coded using a restricted symbol library provided by Airbus. It has a low computational complexity. To verify the scheme robustness against other types of faults, it has also been tested for the control surface liquid and solid jamming (locked in place; see Chap. 4). The results with parametric variations are presented in Fig. 3.17. These results demonstrate that the proposed monitoring scheme is not sensitive to such type of faults.

Finally, the overall approach has been implemented in the flight control computer, coupled with real actuators (ATF; see Sect. 3.3). As an example, during an experiment, the real pilot order is illustrated in Fig. 3.18.

The generated residuals confirmed the satisfactory results obtained from Monte Carlo simulations.

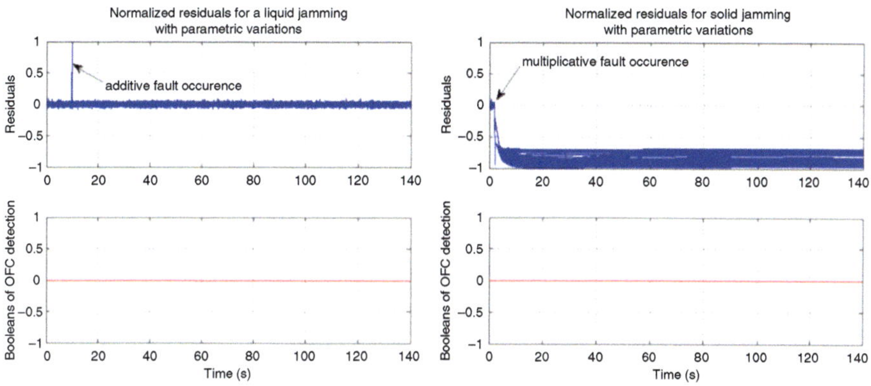

Fig. 3.17 Normalized residual for solid (*right*) and liquid (*left*) jamming with parametric variations

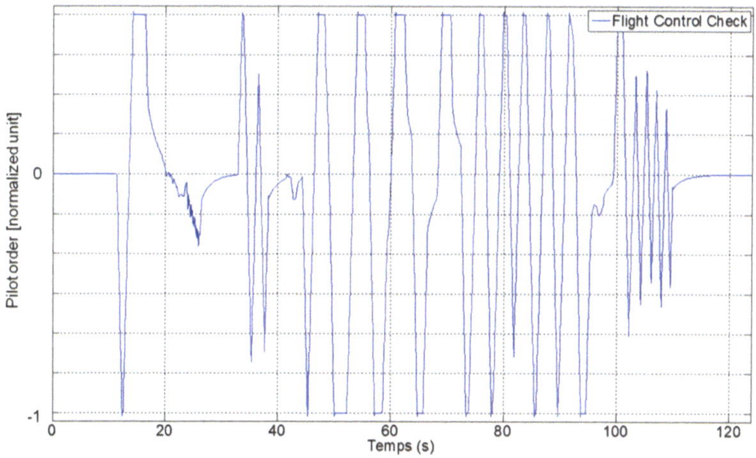

Fig. 3.18 Real pilot order for robustness analysis

3.5 Fault Reconstruction via Sliding-Mode Differentiation

This section presents another appealing approach which is unknown input estimation for OFC reconstruction. The technique to a hybrid robust nonhomogeneous finite-time differentiator, which provides bounded derivatives in noisy environment with a guaranteed accuracy, is presented in [16]. The fault reconstruction is done by solving online a nonlinear equation using a gradient descent method to get a low computational load. This fault reconstruction algorithm is then associated with the same decision-making rules as in the previous section. An advantage of the developed approach is the possibility to build consistency checks directly based on the input signal. In fact, the detection of high-frequency OFCs from output residuals

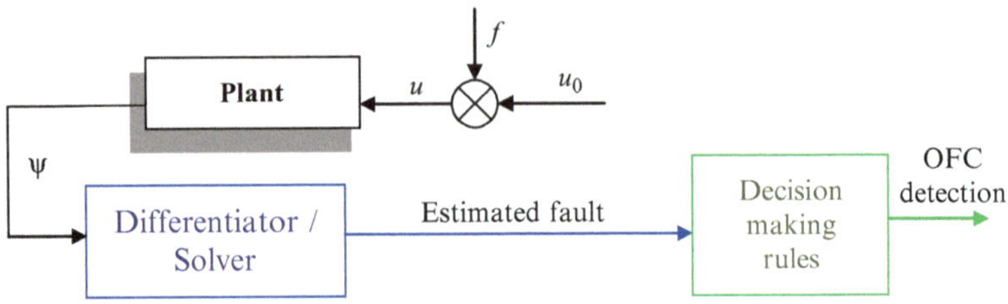

Fig. 3.19 Structure of the hybrid monitoring system

may become difficult as the permissible time window for detection is narrow. The developed scheme has been adapted to satisfy implementation/computation constraints in a FCC (low computational load, restricted symbol library, etc.).

For a better understanding, we give specific notations which will be used in this section: Euclidean norm for a vector $\mathbf{x} \in \mathbb{R}^n$ will be denoted as $|\mathbf{x}|$, and for a measurable and locally essentially bounded input $u : \mathbb{R}_+ \to \mathbb{R}$ ($\mathbb{R}_+ = \{\tau \in \mathbb{R} : \tau \geq 0\}$), the L_∞ norm is denoted as $||u||_{[t_0,T]} = \text{ess sup}\{\,|u(t)|,\ t \in [t_0, T]\}$; if $T = +\infty$ then we will simply write $||u||$. We will denote as $\mathcal{L}_\infty$ the set of all inputs u with property $||u|| < +\infty$. Strictly increasing functions $\sigma : \mathbb{R}_+ \to \mathbb{R}_+$ with the property $\sigma(0) = 0$ form the class $\mathcal{K}$. Recall that the sequence of integers $1, \ldots, k$ is denoted as $\overline{1,k}$.

3.5.1 *Design of Hybrid Differential Observer*

To start, let us take a look at the structure of the monitoring scheme which will be developed in this paragraph. The bloc diagram is given in Fig. 3.19. Its detailed description, the mathematical development, stability proof, accuracy of derivate estimates, fault reconstruction, and OFC detection in noisy environment are presented in the subsequent sections.

3.5.1.1 Differentiator: Boundedness and Accuracy of Derivatives

Consider again the servo-controlled nonlinear actuator model (3.4):

$$\dot{y}(t) = F(t, y, u), \quad \psi(t) = y(t) + v(t) \quad t \geq 0.$$

Here, the measurement noise will be explicitly taken into account: ψ is the measured output and v is the measurement noise, $v \in \mathcal{L}_\infty$. In this case a variant of super-twisting differentiator can be used to provide robust derivative estimate against a non-differentiable noise of any amplitude. Finite-time convergence and accuracy of derivatives can be computed [16]. The sliding-mode differentiator is given by

$$\dot{z}_0 = -\alpha\sqrt{|z_0 - \psi(t)|}\text{sign}[z_0 - \psi(t)] + z_1, \quad \alpha > 0,$$
$$\dot{z}_1 = -\beta\text{sign}[z_0 - \psi(t)] - \chi\text{sign}(z_1) - z_1, \quad \beta > \chi \geq 0, \tag{3.13}$$

where $z_0 \in \mathbb{R}$, $z_1 \in \mathbb{R}$ are the state variables of the system (3.13). α, β, and χ are the tuning parameters. The variable $z_0(t)$ serves as an estimate of the function $y(t)$ and $z_1(t)$ converges to $\dot{y}(t)$. Therefore, (3.13) has the input $\psi(t)$ and the output $z_1(t)$.

The system (3.13) is discontinuous and affected by the disturbance v. Firstly, it should be proved that the system has bounded trajectories. Secondly, we would like to show that the accuracy of derivatives estimation depends continuously on the noise amplitude of v. To proceed, introduce the variables $e_0 = z_0 - y$, $e_1 = z_1 - \dot{y}$, we rewrite the system (3.13) as follows:

$$\dot{e}_0 = -\alpha\sqrt{|e_0|}\text{sign}[e_0] + e_1 + \delta_0(t)$$
$$\dot{e}_1 = -\gamma(t)\text{sign}[e_0] - \chi\text{sign}[e_1] - e_1 + \delta_1(t)$$
$$\delta_0(t) = \alpha\left(\sqrt{|e_0|}\text{sign}[e_0] - \sqrt{|e_0 - v(t)|}\text{sign}(e_0 - v(t))\right)$$
$$\delta_1(t) = \beta\left(\text{sign}(e_0) - \text{sign}(e_0 - v(t))\right) \tag{3.14}$$

where δ_0, δ_1 are the disturbances originated by the noise v presence, $\gamma(t) = \beta + (\dot{y}(t) + \ddot{y}(t) - \chi(\text{sign}[e_1(t)] - \text{sign}[e_1(t) + \dot{y}(t)]))\text{sign}[e_0(t)]$ is a piecewise continuous function (for $\beta > l_1 + l_2 + 2\chi$ it is strictly positive and $0 < \delta \leq \gamma(t) \leq \kappa$, $\delta = \beta - l_1 - l_2 - 2\chi$, $\kappa = \beta + l_1 + l_2 + 2\chi$). Assume that $|v(t)| \leq \lambda_0$ for all $t \in \mathbb{R}_+$. By definition $|\delta_0(t)| \leq \alpha\sqrt{2\lambda_0}$, $\delta_1(t) = 0$ for $|e_0(t)| \geq \lambda_0$, $|\delta_1(t)| \leq 2\beta$, and $\delta_1(t)e_0(t) \geq 0$ for all $t \in \mathbb{R}_+$. Then the global boundedness of solutions of the differentiator (3.14) is proven in the next lemma.

Lemma 3.3 *Let the signal* $v : \mathbb{R} \to \mathbb{R}$ *be Lebesgue measurable and* $|\dot{y}(t)| \leq l_1$, $|\ddot{y}(t)| \leq l_2$, $|v(t)| \leq \lambda_0$ *for all* $t \in \mathbb{R}_+$*;* $\alpha > 0$, $\beta > 0$, *and* $0 < \chi < \beta$. *Then in* (3.14) *for all* $t_0 \in \mathbb{R}_+$ *and initial conditions* $z_0(t_0) \in \mathbb{R}$, $z_1(t_0) \in \mathbb{R}$ *the solutions are bounded:*

$$|z_0(t) - y(t)| < \max\left\{|z_0(t_0) - y(t_0)|, 4\alpha^{-2}(|z_1(t_0) - \dot{y}(t_0)| + 3\beta + l_1 + l_2 + \chi + \alpha\sqrt{2\lambda_0})^2\right\}, |z_1(t) - \dot{y}(t)| \leq |z_1(t_0) - \dot{y}(t_0)|\,\mathrm{e}^{-0.5t} + |3\beta + l_1 + l_2 + \chi|.$$

Proof Let us start with the second equation in the system (3.14), considering the Lyapunov function $U(e_1) = 0.5{e_1}^2$, which has the derivative $\dot{U} \leq -U + 0.5[3\beta + l_1 + l_2 + \chi]^2$. That gives the desired estimate. Next consider $U(e_0) = 0.5{e_0}^2$, then $\dot{U} \leq -\alpha\sqrt{|e_0|}\,|e_0| + |e_0|\,(|e_1| + \alpha\sqrt{2\lambda_0})$, and since $|e_1(t)| \leq |e_1(t_0)| + 3\beta + l_1 + l_2 + \chi$ for $|e_1(t_0)| + 3\beta + l_1 + l_2 + \chi + \alpha\sqrt{2\lambda_0} \leq 0.5\alpha\sqrt{|e_0|}$, we have $\dot{U} \leq -0.5\alpha\sqrt{|e_0|}\,|e_0| < 0$ that implies Lemma.

Let us analyze now the accuracy of derivatives in the presence of a non-differentiable noise; let the signal $v : \mathbb{R} \to \mathbb{R}$ be Lebesgue measurable and $|v(t)| \le \lambda_0$ for all $t \in \mathbb{R}_+$.

Theorem 3.3 *Let* $\beta > l_1 + l_2 + 2\chi$, $\chi > 0$, *and* $\alpha \ge 2\{\sqrt{8\kappa}\chi + \sqrt{\chi+\kappa}(\kappa-\delta)\}/(1.5\delta + 0.5\kappa)$, *then for any initial conditions* $\mathbf{e}(0) \in \Omega_0$, $\Omega_0 = \left\{\mathbf{e} \in \mathbb{R}^2 : \kappa\,|e_0(0)| + 0.5{e_1}^2(0) \le 2\sqrt{2\kappa(\chi+\kappa)}\chi\delta(\kappa-\delta)^{-1}\right\}$ *the trajectories of the system* (3.14) *satisfy the estimate for all* $t \ge T$

$$|e_0(t)| \le \delta^{-1}(c_1\lambda_0 + c_2\sqrt{\lambda_0}),\ |e_1(t)| \le \sqrt{2(c_1\lambda_0 + c_2\sqrt{\lambda_0})},\ c_2 = \beta^2/(\alpha\sqrt{2}),$$

$$c_1 = \max\{8\mu^{-2}[(0.25\delta+\kappa)\alpha + \max\{\sqrt{2(\chi+\kappa)}, 6\}\beta]^2, \kappa\},\ \mu = \min\{\alpha\delta/\sqrt{\kappa}, \sqrt{2}\chi\},$$

where the finite time T *of convergence possesses the estimate* $T \le 4\mu^{-1}\sqrt{\kappa\,|e_0(0)| + 0.5{e_1}^2(0)}$, *provided that*

$$c_1\lambda_0 + c_2\sqrt{\lambda_0} \le 2\sqrt{2\kappa(\chi+\kappa)}\,\chi\delta(\kappa-\delta)^{-1}.$$

Proof See [16].

The Theorem 3.3 is based on the observation that δ_1 (the product $e_1\delta_1$) influences negatively onto the set $\Gamma = \left\{|e_0| < \lambda_0 \wedge 3\alpha\sqrt{2\lambda_0} < |e_1| < 2\beta \wedge e_0 e_1 > 0\right\}$ only. The result of the theorem says that if the noise amplitude λ_0 is comparable with the chosen α, β, χ (the constraint $c_1\lambda_0 + c_2\sqrt{\lambda_0} \le 2\sqrt{2\kappa(\chi+\kappa)} \times \chi\delta(\kappa-\delta)^{-1}$ holds), then the estimate on the derivative $\dot{y}$ has the error proportional to $\lambda_0^{0.25}$. If the noise amplitude is very high, then the result of Lemma 3.3 is satisfied guaranteeing boundedness of trajectories. It is worth to stress that the value $2\sqrt{2\kappa(\chi+\kappa)} \times \chi\delta(\kappa-\delta)^{-1}$ can be taken arbitrarily high adjusting α, β, χ.

3.5.1.2 Fault Reconstruction

The estimation algorithm design for the fault signal f reconstruction is performed in two steps in this subsection. Firstly, the main assumptions are introduced. Secondly, a hybrid algorithm is presented, and its conditions of convergence and accuracy are analyzed.

Assume that the state of (3.4) belongs to some (may be unknown) compact set.

Assumption 3.3 *Let* $(y(t), \dot{y}(t)) \in Y \subset \mathbb{R}^2$ *for all* $t \ge 0$.

This assumption is quite realistic. Typically, the set Y is known and predefined during the design phase. When the faults f are present, the system (3.4) may lose its stability. However, as it will be shown below even in this case, the algorithm requires a finite time to detect the fault. Hence, recovery actions can be made to maintain stability and some predefined performance level. In addition, for the actuator model (3.4) this assumption is clearly satisfied due to physical constraints (the system (3.4) is stable with a bounded input u, which is the case with or without the fault f). Thus, compact sets Y and U can be computed (taking a priori information available about conditions of operation of a particular actuator) such that $y(t) \in Y$ and $u(t) \in U$ for all $t \geq 0$.

According to Theorem 3.3, for the system (3.13), there exists a finite time of convergence $T \leq 4\mu^{-1}\sqrt{\kappa\,|e_0(0)| + 0.5{e_1}^2(0)}$ such that $y^{(k)} = z_k(t) - e_k(t)$, $k = 0, 1$ with $|e_0(t)| \leq \delta^{-1}(c_1\lambda_0 + c_2\sqrt{\lambda_0})$ and $|e_1(t)| \leq \sqrt{2(c_1\lambda_0 + c_2\sqrt{\lambda_0})}$ for all $t \geq T$. Then, the system (3.4) can be presented as follows:

$$z_1(t) - e_1(t) = F(t, z_0(t) - e_0(t), u(t)), \quad t \geq 0. \tag{3.15}$$

Let $Y_\upsilon \subset \mathbb{R}^2$ be the neighborhood of the set Y ($Y \subset Y_\nu$) such that if $|z_0(t) - y(t)| \leq \delta^{-1}(c_1\lambda_0 + c_2\sqrt{\lambda_0})$, $|z_1(t) - y'(t)| \leq \sqrt{2(c_1\lambda_0 + c_2\sqrt{\lambda_0})}$, and $(y, y') \in Y$, then necessarily $(z_0, z_1) \in Y_\upsilon$. Since the function F is locally Lipschitz continuous, then for all $(z_0, z_1) \in Y_\upsilon$ there exists $L > 0$ such that

$$|F(t, z_0(t) - e_0(t), u(t)) - F(t, z_0(t), u(t))| \leq L\,|e_0(t)|\,.$$

According to Theorem 3.3 we have $|e_0(t)| \leq \delta^{-1}(c_1\lambda_0 + c_2\sqrt{\lambda_0})$ for all $t \geq T$. Therefore, from the expression (3.15) we can define the augmented error

$$\begin{aligned}\delta(t) &= z_1(t) - F(t, z_0(t), u_0(t) + f(t)) \\ &= F(t, z_0(t) - e_0(t), u(t) + f(t)) + e_1(t) - F(t, z_0(t), u_0(t) + f(t))\end{aligned} \tag{3.16}$$

with $|\delta(t)| \leq \rho(||v||)$ for all $t \geq T$, $\rho(s) = L\delta^{-1}(c_1 s + c_2\sqrt{s}) + \sqrt{2(c_1 s + c_2\sqrt{s})}$.

All variables in the right-hand side of (3.16) are available for measurements except the fault signal $f(t)$. In the left-hand side of (3.16) we have the augmented error δ, which represents the accuracy of the derivative estimation by the differentiator (3.13); it is not measurable and it is proportional to the measurement noise v amplitude (this error becomes zero in the finite time T for the case of no measurement noise). Let $\widehat{f}(t)$ be a solution of the Eq. (3.16) for the case $\delta(t) = 0$, i.e.,

$$z_1(t) = F(t, z_0(t), u_0(t) + \widehat{f}(t)), \tag{3.17}$$

then substituting (3.17) in (3.16) we get

$$\delta(t) = F(t, z_0(t), u_0(t) + \widehat{f}(t)) - F(t, z_0(t), u_0(t) + f(t)).$$

Define the gradient of the function F with respect to u

$$\nabla_u F(t, y, u) = \partial F(t, y, u)/\partial u,$$

then by the mean value theorem (see Theorem 3.1), there exists a function $c : \mathbb{R}_+ \to [0, 1]$ such that for all $t \geq 0$

$$\begin{aligned} \delta(t) &= g(t)[\widehat{f}(t) - f(t)], \\ g(t) &= \nabla_u F(t, z_0(t), u_0(t) + [1 - c(t)]\widehat{f}(t) + c(t)f(t)). \end{aligned} \tag{3.18}$$

Assumption 3.4 *Let*

$$\int_0^t |g(\tau)[\widehat{f}(\tau) - f(\tau)]| d\tau \geq g_{\min} t |\widehat{f}(t) - f(t)|^p$$

for all $t \geq T$ *and some* $g_{\min} > 0$, $0 < p < +\infty$.

The condition of Assumption 3.4 means that on the time interval $t \geq T$ the integral $\int_0^t |g(\tau)^T[\widehat{f}(\tau) - f(\tau)]| d\tau$ has average value bigger than $g_{\min}|\widehat{f}(t) - f(t)|^p$. Roughly speaking this property says that the function $g : \mathbb{R}_+ \to \mathbb{R}$ norm has a strictly separated from zero average value for all $t \geq T$. This property can also be considered as a variant of the well-known persistency of excitation condition in the estimation/adaptation theory [17]. Then under Assumption 3.4 from (3.18) for all $t \geq T$, we obtain the upper estimate

$$\rho(\|v\|)t \geq \int_0^t |\delta(\tau)| \, d\tau = \int_0^t \left| g(\tau)[\widehat{f}(\tau) - f(\tau)] \right| d\tau \geq g_{\min} t \left| \widehat{f}(t) - f(t) \right|^p$$

and, finally,

$$|\widehat{f}(t) - f(t)| \leq [g_{\min}^{-1} \rho(||v||)]^{1/p},$$

which implies boundedness of the discrepancy $\widehat{f}(t) - f(t)$ for all $t \geq T$. In other words, accuracy of the fault signal f estimation by $\widehat{f}$ is a function of the measurement noise v amplitude. Consequently, under Assumption 3.4 the problem of fault detection and isolation can be handled finding a solution $\widehat{f}$ of the Eq. (3.17), the penalty is proportional to $||v||$.

The Eq. (3.17) is nonlinear; for each $t \geq 0$ it may have a single solution $\widehat{f}(t)$ or in general case, $\widehat{f}(t) \in S_t$, where for all elements $s \in S_t$ the equation $z_1(t) = F(t, z_0(t), u_0(t) + s)$ holds. It could be the case that for some $t \geq 0$, this equation has no solution with respect to $\widehat{f}(t)$. Thus, some regularizing conditions have to be imposed.

Assumption 3.5 *Let* $\nabla_u F(t, z_0(t), u(t)) \neq 0$ *for all* $(z_0, z_1) \in Y_\upsilon$, $u \in \mathbb{R}$, *and* $t \geq 0$.

Note that Assumption 3.5 does not necessarily imply Assumption 3.4. This assumption states that the gradient of the function F with respect to the last argument u is restricted from zero, or in other words, under these restrictions the Eq. (3.17) has the single solution $\widehat{f}(t)$. To verify this assumption, let us compute

$$\nabla_u F(t, y, u) = \frac{K_{\rm ci} K \Delta P_{\rm ref}}{\Delta P_{\rm ref} + K_a(t)[K_{\rm ci} K(u-y)]^2} \sqrt{\frac{\Delta P(t) - S^{-1} F_{\rm aero}(t)}{\Delta P_{\rm ref} + K_a(t)[K_{\rm ci} K(u-y)]^2}},$$

it is easy see that $\nabla_u F(t, y, u) > 0$ for any finite $u \in U$ and $y \in Y$, and the assumption is satisfied. To verify Assumption 3.4, compute

$$g(t) = \nabla_u F\{t, \psi(t), u_0(t) + [1 - c(t)]\widehat{f}(t) + c(t) f(t)\},$$

which implies $g(t) \geq g_{\min} > 0$ due to the form of $\nabla_u F$ presented above, for any $u_0 \in U$, $y \in Y$ and a finite harmonic fault f with its estimate $\widehat{f}$. Then this assumption is also valid.

Under Assumption 3.5, any gradient descent method can be applied to find an estimate $\breve{f}(t)$ on the solution of (3.17) $\widehat{f}(t)$:

$$\begin{aligned} &\mathrm{d}\breve{f}(\tau)/d\tau = \gamma\phi[\sigma(t, \breve{f}(\tau))], \\ &\varphi(s)s > 0 \text{ for all } s \neq 0,\ ||\varphi|| < +\infty, \\ &\sigma(t, f) = [z_1(t) - F(t, z_0(t), u_0(t) + f)]\nabla_u F(t, z_0(t), u_0(t) + f) \end{aligned} \tag{3.19}$$

where $\gamma > 0$ is a design parameter and $\tau \geq 0$ is an independent time. For each fixed $t \geq 0$, the execution of (3.19) in the time τ ensures convergence of $\breve{f}(\tau)$ to $\widehat{f}(t)$ (morc precisely this claim will be formulated later).

Under the introduced Assumptions 3.3–3.5, the proposed fault isolation algorithm consists in discretization of (3.19), when the estimate $\breve{f}(t_k)$ is generated discretely for some sequence of strictly increasing sample instants t_k, $k \geq 0$ $(t_0 = 0)$ having accumulation point at infinity only. Then, the discrete representation of (3.19) can be written as follows for any $k \geq 0$:

$$\theta_0 = \breve{f}(t_k), \breve{f}(t_0) = \breve{f}_0; \theta_{r+1} = \theta_r + \gamma\phi[\sigma(t_k, \theta_r)], r = \overline{0, N-1}; \breve{f}(t_{k+1}) = \theta_N \tag{3.20}$$

where $\gamma > 0$, $N > 0$, and $\breve{f}_0 \in \mathbb{R}$ are design parameters. The operation of (3.20) can be explained as follows: at each sampling time t_k the algorithm takes the initial value $\theta_0 = \breve{f}(t_k)$ (or some guess $\theta_0 = \breve{f}_0$ on the first step $k = 0$), then N steps of the discrete minimization procedure (3.20) are computed, and the output of the algorithm is $\breve{f}(t_{k+1}) = \theta_N$. The number N is bounded by available computational power for (3.20) realization. The system (3.20) period or the shift between the sample instants $t_{k+1} - t_k$, $k \geq 0$ depends on the time that is required to perform N steps of (3.20) and the fault detection minimum time specifications. The stability properties of the obtained hybrid system (3.4), (3.13), and (3.20) (its structure scheme is shown in Fig. 3.16) are analyzed below.

Theorem 3.4 *Let Assumptions* 3.3–3.5 *hold, then in the system* (3.4), (3.13), and (3.20) *for any* $\varepsilon^* > 0$ *there exist* $\gamma^* > 0$ *and* $N^* \geq 0$ *such that for any* $k > 0$ *with* $t_k \geq T$ *(where* $T \geq 0$ *is the time of the derivatives estimation from Theorem* 3.2)

$$|\breve{f}(t_{k+1}) - f(t_k)| \leq \varepsilon^* + [g_{\min}^{-1}\rho(||v||)]^{1/p}$$

provided that $0 < \gamma < \gamma^*$, $N \geq N^*$ *for any initial conditions,* $v \in \mathcal{L}_\infty$ *and continuous* $f \in \mathcal{L}_\infty$.

Proof See [18].

The result of Theorem 3.4 claims that for any desired accuracy $\varepsilon^* > 0$ there exists some maximum adaptation rate $\gamma^* > 0$ and maximum number of steps $N^* \geq 0$ such that the fault value $f(t_k)$, $t_k \geq T$ for all such $k \geq 0$ is estimated by the algorithm (3.20) output $\breve{f}(t_{k+1})$ with the worst-case accuracy $[g_{\min}^{-1}\rho(||v||)]^{1/p} + \varepsilon^*$. In the absence of the measurement noise v the accuracy ε^* is achievable. The theorem does not restrict the sampling rate in the system (the delay $t_{k+1} - t_k$, $k \geq 0$ can be chosen in accordance with computational constraints). There exists a casual time shift in the algorithm response ($\breve{f}(t_{k+1}) \to f(t_k)$) due to calculations in (3.20) performed on the interval $[t_k, t_{k+1})$; the estimate on the value $f(t_k)$ is always obtained on the next step t_{k+1} only.

In particular, for FDD purposes, if $0 < t_{k+1} - t_k \leq T_0$ ($T_0 > 0$ is the maximal sample time of the algorithm (3.20) operation), then Theorem 3.4 guarantees that for time instants $t_k \geq T + T_0$, $k \geq 0$ the signal $\breve{f}(t_k)$ detects all faults with amplitudes bigger than $[g_{\min}^{-1}\rho(||v||)]^{1/p} + \varepsilon^*$ (in other words, $T + T_0$ is the fault detection time and $[g_{\min}^{-1}\rho(||v||)]^{1/p} + \varepsilon^*$ represents the amplitude of the smallest detectable fault).

3.5.2 Experimental Results

In Sect. 3.4, OFCs occurring in servo-controlled elevator surfaces have been considered. In this section, the considered OFC are those related to the right inboard aileron. The requirement specifications are the following: 0 % of missed detection, 0 % of false alarm, and 100 % of true detection for all flight conditions. In the following case study, the detection delay requirement is to detect an OFC in less than 3 cycles. Recall that, as described in Sect. 3.2, the OFC decision-making rule used at Airbus consists in counting successive and alternate crossings of a given threshold η in a sliding time window. In the case of liquid failures, the residual is given by

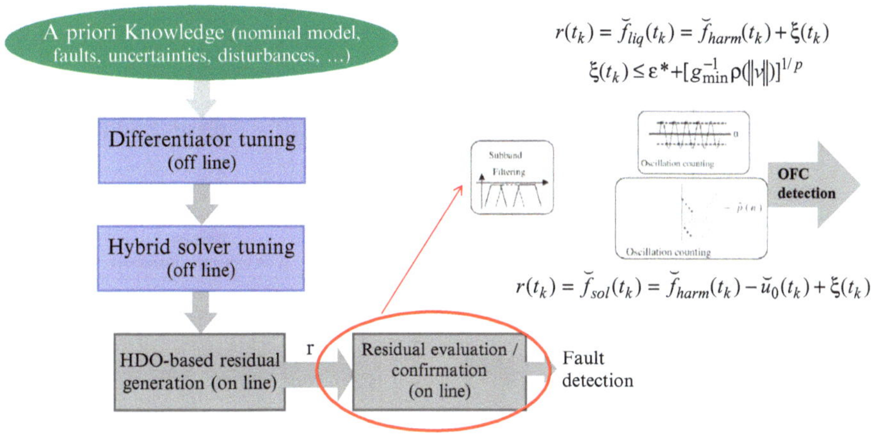

Fig. 3.20 Bloc diagram for OFC detection

$$r(t_k) = \breve{f}_{\text{liq}}(t_k) = \breve{f}_{\text{harm}}(t_k) + \xi(t_k),$$

where $|\xi(t_k)| \leq \left|\varepsilon^* + [g_{\min}^{-1}\rho(\|v\|)]^{1/p}\right|$. After a filtering, the residual is zero averaged, and OFC can be detected by counting around zero alternate and successive crossings of a threshold (see Sect. 3.2).

In the case of a solid failure, the OFC substitutes the nominal signal, and then the residual is expressed as

$$r(t_k) = \breve{f}_{\text{sol}}(t_k) = \breve{f}_{\text{harm}}(t_k) - u_0(t_k) + \xi(t_k).$$

If the control position u_0 is null (no control surface deflection), then the residual is only composed of a failure and bounded error ξ. OFC detection can be thus done by oscillation counting around zero, like for a liquid failure. However, if a control surface deflection is required by the flight control law (e.g., during a maneuver or in reaction to the failure), the failure signal is mixed with the opposite of the estimated position, and an oscillation counting around zero would not enable detection. In this case, it is proposed to count OFC on the residual signal but around the opposite of the estimated position. Note that both liquid and solid OFC decision blocks operate in parallel. A structure scheme of OFC detection system is shown in Fig. 3.20. Finally, note that if the FDD is supposed to detect other kinds of failure in the actuator control loop, the decision rule could be just a simple threshold-based logic. However, the goal here is to detect, and further to confirm, that the fault to be detected is an OFC, and not something else. That is why the above evaluation rule from [2] is used.

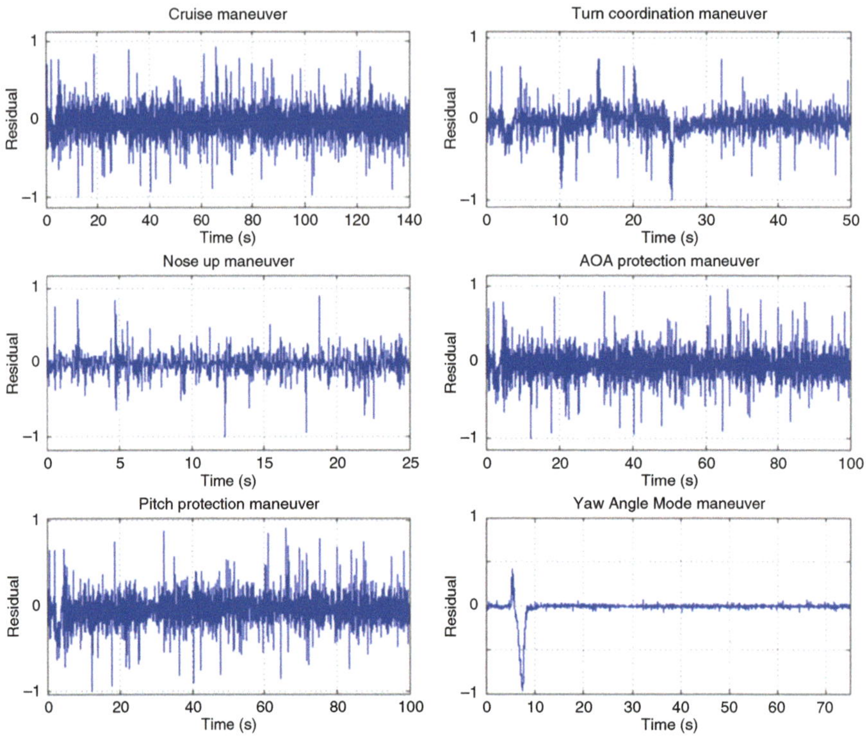

Fig. 3.21 Robustness analysis – normalized residuals on right inboard aileron

The actuator modeling is based on three elements: the actuator model itself, a control surface position saturation that could be dissymmetric, and a rate limiter representing the physical limitations.

3.5.2.1 Airbus Aircraft Benchmark Results

To assess the potential of this FDD scheme, the six different flight maneuvers as used in Sect. 3.4.4 are simulated for fault-free situations. The aforementioned flight scenarios are used to show the good robustness and performance of FDD scheme for both lateral and longitudinal modes. Firstly, the robustness of the proposed FDD scheme is studied. The results of benchmark simulations (AAB) show that the amplitude of residual stays small for five flight maneuvers (see Fig. 3.21). For the yaw angle mode scenario, a dynamic phase introduces some vibrating behavior in the residual due to an important variation in the aerodynamic forces. Since it is necessary to have a limited number of successive oscillations to detect a fault, the monitoring algorithm concludes well to a nominal (no fault) situation. Hence, the proposed residual generation gives no false alarm. Secondly, the fault detection ability is checked. For this purpose, several OFC amplitudes and frequencies are tested for both liquid and solid OFC. In all cases, the oscillatory phenomenon appears clearly on residual signal, and the fault is detected by the decision block. It follows that for all tests, we get 100 % of true detection and 0 % of missed detection.

3.5.2.2 FES Parametric Simulation Results

According to the tests performed in Sect. 3.4.4, several simulations have been conducted for the six previous flight scenarios in fault-free situations with variations in the operating conditions and uncertainties (see Sect. 3.4.4 for more details). Figure 3.22 shows the residuals obtained in FES environment for healthy situations where no fault is detected. The results show a good robustness against parametric variations since the FOM gives 0 % of false alarms for the 1,200 realistic fault-free simulation runs. In addition, note that the parametric tests involve some unwanted oscillatory behaviors of residuals between 0 and 5 s (see the yaw angle mode maneuver in Fig. 3.22). These behaviors are due to the command signal generated by the flight control unit. On the other hand, some normal oscillations with the frequency between 0.1 and 1 Hz can be observed in addition to the faults to be detected, making the detection of OFC more difficult. Hence, the detection threshold of the decision block has to be set to a higher value to keep 0 % of false alarm for all situations.

Next, the faulty situations with the parametric variations have been analyzed for the angle of attack protection maneuver. Due to an important number of data generated during the simulations, only the results corresponding to the smallest OFC amplitude and the minimal and maximal OFC frequencies (0.5 and 7 Hz, respectively) are given. Figure 3.23 shows the normalized residuals for liquid (left part) and solid (right part) faults, respectively. As it can be seen, the residual is small before the fault occurrence for both liquid and solid faulty situations. Next, a significant change of the residual appears. For the liquid faults, the residual is a noisy sinusoid where the frequency of this sinusoid coincides with the frequency of the fault (see Fig. 3.23). In all cases, there is no missed detection. The statistical results given by FOM for the smallest OFC amplitude (Airbus specifications not given here for industrial reasons) are summarized in Table 3.1. The detection time (DT) index is used to quantify the detection delay requirement in a normalized way, i.e., $DTP < 1$ denotes an enhancement of detection delay, $1 < DTP < 1.3$ represents an acceptable level of performances, and $DTP > 1.3$ is judged like unacceptable detection delays. From Table 3.1, OFCs are always detected with satisfactory detection time, i.e., the proposed hybrid monitoring scheme obtains 100 % of true detection and 0 % of missed detection for the considered flight maneuvers with acceptable detection delays.

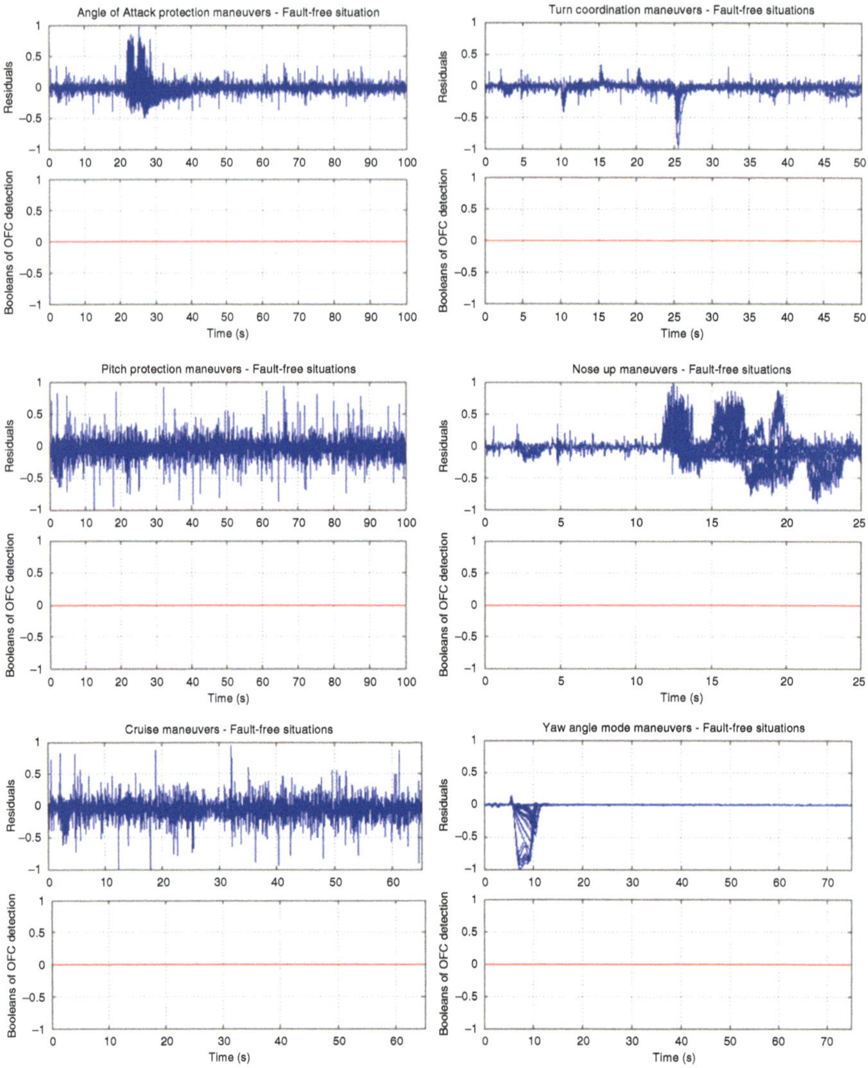

Fig. 3.22 Normalized residuals in fault-free situations with parametric variations: Fault reconstruction via sliding-mode differentiation

3.5.2.3 Implementation Aspects

The proposed scheme has been coded using an in-house restricted symbol library used by Airbus [8, 9] for real-time FCC software implementation. The low computational complexity of the proposed detection method is confirmed by the fact that it needs 322 basic operators only (like delays, multiplications, additions, gains, sign function, look-up tables, logic operators). The computational load can

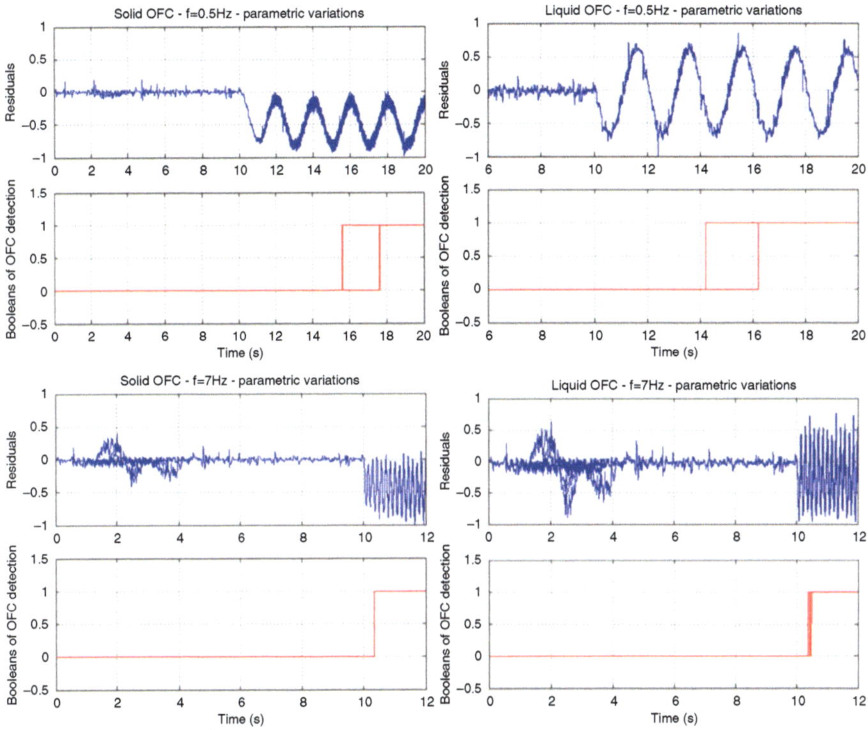

Fig. 3.23 Normalized residuals for liquid (*left*) and solid (*right*) fault

Table 3.1 FOM for OFC fault in the right inboard aileron – parametric tests

			Normalized detection time (DT)				
Type	Amp	f(Hz)	Mean	Max	Min	True det. (%)	Missed det. (%)
Liquid	Smallest	0.5	0.747	1.03	0.7	100	0
Solid	Smallest	0.5	1.06	1.27	0.93	100	0
Liquid	Smallest	7	0.9471	1.12	0.84	100	0
Solid	Smallest	7	0.79	0.79	0.79	100	0

be evaluated by using the running time of each symbol. It follows that the proposed strategy uses only 47 % of computing cost allowed in the ADDSAFE project for OFC detection. Another interesting feature of this FDD scheme deals with the robustness against other types of fault. In this case, Fig. 3.24 shows the results of control surface liquid and solid jamming with parametric variations (see Chap. 4 for this fault analysis). Simulation results confirm that the proposed monitoring scheme is not sensitive to such type of faults. It is a great feature of the considered approach since a simple threshold-based logic will conclude to fault detection, i.e., there is no robustness against liquid and solid jamming.

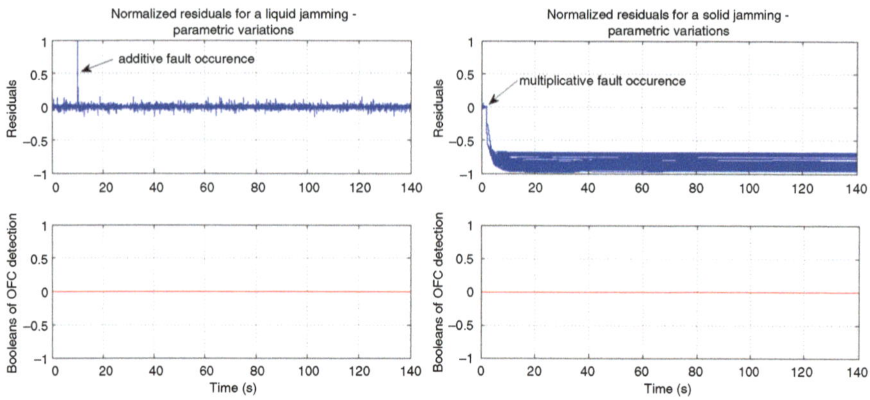

Fig. 3.24 Normalized residual for solid (*right*) and liquid (*left*) jamming with parametric variations

3.6 Conclusion

This chapter studied the problem of oscillatory fault detection in the actuators of a modern civil aircraft. A reliable detection of this kind of fault is very important for early system reconfiguration and for structural design optimization toward a greener aircraft. After presenting the industrial state-of-practice solution used on in-service A380 aircraft, two FDD schemes are analyzed, a nonlinear observer and a hybrid monitoring scheme based on a sliding-mode differentiator. In all cases, validation results using intensive Monte Carlo simulations have been presented. During summer 2009, the first approach was successfully implemented and tested on an A380 simulator[4] (Airbus, Toulouse, France). The second approach has been tested and validated through Monte Carlo campaigns during ADDSAFE project (see Chap. 1).

References

1. Stengel RF (2004) Flight dynamics. Princeton University Press, Princeton
2. Goupil P (2010) Oscillatory failure case detection in the A380 electrical flight control system by analytical redundancy. Control Eng Pract 18(9):1110–1119
3. Alcorta-Garcia E, Zolghadri A, Goupil P (2011) A nonlinear observer-based strategy for aircraft oscillatory failure detection: A380 case study. IEEE Trans Aerosp Electron Syst 47(4):2792–2806

[4]http://www.ims-bordeaux.fr/IMS/ressources/fichiers/NGE2YzJjYjY2OTAyZg==/A380_anglais.pdf

4. Trapero JR, Sira-Ramirez H, Feliu Batlle V (2007) An algebraic frequency estimator for a biased and noise sinusoidal signals. Signal Process 87:1188–1201
5. Hou M (2007) Estimation of sinusoidal frequencies and amplitudes using adaptive identifier and observer. IEEE Trans Autom Control 52(3):493–499
6. Loutridis SJ (2004) Damage detection in gear systems using empirical mode decomposition. Eng Struct 26(12):1833–1841
7. Giesseler HG, Besch HM (1995) The oscillatory failure identification system (OFIS). In: Proceedings of the 36th AIAAS/ASME/ASCE/AHS/ASC structures, structural dynamics and material conference, New Orleans, LA pp 3304–3317
8. Goupil P, Puyou G (2011) A high fidelity AIRBUS benchmark for system fault detection and isolation and flight control law clearance. In: Proceedings of the 4th European conference for aerospace sciences, Saint-Petersburg
9. Goupil P, Marcos A (2012) Industrial benchmarking and evaluation of ADDSAFE FDD designs. In: Proceedings of the 8th IFAC symposium SAFEPROCESS, Mexico
10. Goupil P, Marcos A (2011) Advanced diagnosis for sustainable flight guidance and control: The European ADDSAFE project. SAE technical paper, 2011-01-2804
11. Fernandez V, De Zaiacomo G, Mafficini A, Peñín LF (2010) The IXV GNC functional engineering simulator, 11th international workshop on simulation & EGSE facilities for space programmes, ESA-ESTEC
12. Frank PM (1978) Introduction to sensitivity theory. Academic Press, New York
13. Thau FE (1973) Observing the state of non-linear dynamic systems. Int J Control 17(3):471–479
14. Apostol TM (1969) Calculus II: multi variable calculus and linear algebra with application to differential and probability, vol II, 2nd edn. Wiley, Hoboken
15. Khalil HK (1996) Nonlinear systems, 2nd edn. Prentice Hall, Englewood Cliffs
16. Efimov D, Fridman L (2011) A hybrid robust non-homogeneous finite-time differentiator. IEEE Trans Autom Control 56(5):1213–1219
17. Narendra KS, Annaswamy AM (1989) Stable adaptive systems. Prentice-Hall, Inc., Englewood Cliffs
18. Efimov D, Zolghadri A, Raïssi T (2011) Actuators fault detection and compensation under feedback control. Automatica 47:1699–1705

Chapter 4
Robust Detection of Abnormal Aircraft Control Surface Position for Early System Reconfiguration

4.1 Introduction

This chapter follows the basic problem addressed in the previous chapter and deals with two other important EFCS-failure cases: runaway (aka hardover) and jamming (or lock-in-place failure) of aircraft control surfaces. Early and robust detection of such failures is also an important issue for achieving sustainability goals and for early system reconfiguration [1]. The chapter focuses on the elevator runaway and jamming. As outlined in the previous chapter, the elevator setting controls the pitch angle, an important function especially during takeoff and landing.

A runaway is an unwanted, or uncontrolled, control surface deflection that can go until the moving surface stops if it remains undetected. Runaway can have various dynamic profiles and is mainly due to an electronic component failure, mechanical breakage, or FCC malfunctions. Low-speed runaway results in an undesired pitch maneuver that may significantly degrade the aircraft controllability and that may increase the pilot workload. High-speed runaways generally do not impact the aircraft trajectory but lead to additional loads that must be taken into account in the aircraft structural design objectives. In any cases, the detection of the runaway must be accomplished before the elevator position exceeds a few degrees from its trimmed value. This aim results in a required detection time depending on the actual fault rate. Common civil aircraft configurations have two independent elevators, each controlled by dual actuators (one active, one in standby), and each actuator controlled by a dedicated FCC, to ensure the aircraft's maneuverability in the case of actuator failures. A detected runaway will first result in servo-control deactivation and then in system reconfiguration which means that there is a hand over between both actuators and between FCC. The active actuator is automatically converted in passive mode, and the passive one is switched in active mode. When the system reconfiguration is triggered, the control surface has already reached a given position (Fig. 4.1). For the design objectives related to structural loads, it is required to

A. Zolghadri et al., *Fault Diagnosis and Fault-Tolerant Control and Guidance for Aerospace Vehicles: From Theory to Application*, Advances in Industrial Control,
DOI 10.1007/978-1-4471-5313-9_4, © Springer-Verlag London 2014

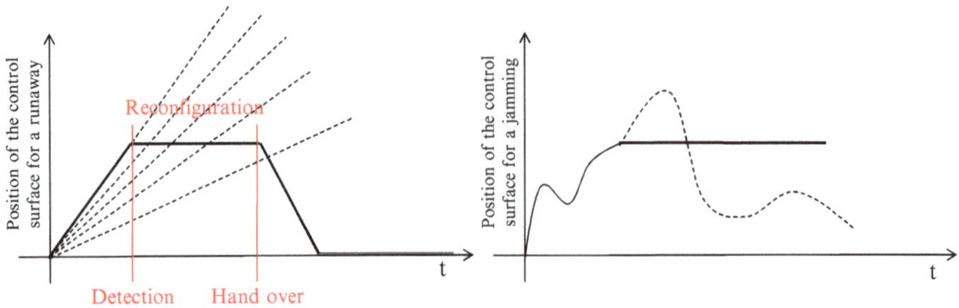

Fig. 4.1 Control surface position for runaway and jamming case

detect very quickly the runaway, before excessive loads appear on the structure. For aircraft controllability, it is also required to detect rapidly the runaway, before a too important trajectory change.

A jamming, or lock-in-place failure, is a generic system-failure case which generates control surface stuck at its current position (Fig. 4.1). The jamming of an aircraft control surface creates a dissymmetry in the aircraft configuration, which must be compensated by appropriate defections of other control surfaces. A well-known negative effect of jamming is the resulting increased drag, which leads to increased fuel consumption since the remaining safe control surfaces stay permanently deflected. Increased fuel burn means an increased environmental footprint and a possible aircraft diversion in case of lack of fuel. For example, during a coordinated turn, if an elevator is jammed, the reaction of the aircraft is weaker, and for compensating, more deflection will be demanded on the remaining elevator as well as on the Trimmable Horizontal Stabilizer. Due to the coupling with the roll axis, an additional dissymmetrical deflection of the aileron will be required. In case of landing with strong crosswind, a stuck rudder could prevent to correctly control the aircraft and to compensate the induced sideslip. Another example is when jamming occurs during a long-time aircraft operation. In this case, a surface jamming may produce substantial drag and again excessive fuel consumption and can even obstruct the fulfillment of the flight mission (i.e., the need for landing on a diverting airport for refueling). Therefore, the timely detection of jamming, especially of the primary control surfaces (e.g., elevator, rudder, ailerons), is important for both economical and easy-to-handle operation of an aircraft.

In this chapter, a simple model-based solution is presented to address the above problems. The technique is based on estimation of the fault model parameters to detect abnormal changes affecting normal operation of the system. Contrarily to the techniques presented in the previous chapter, here, for the runaway case, the system model is not used for developing FDD algorithm. As it will be shown, the approach has very attractive features from a practical point of view: as no system model (local or global) is used, the fault model is independent of system model. Another important advantage is that the method can easily be adapted to any control surface on any aircraft model. The validation of the detection system is also independent of all possible system parametric variations. In fact, the deterministic part of the fault model does not depend on the system parameters, and its stochastic part is used to manage the performance/robustness trade-offs by adjusting the tuning parameters. It will be shown that by careful fault modeling, simple estimation techniques can lead to remarkable results. Finally, a big advantage of the proposed scheme is that it can be embedded within the structure of in-service monitoring system as a part of the FCC software. The proposed technique is based on a combined data-driven and model-based approach using a dedicated Kalman filtering, providing an effective method to achieve well-defined real-time characteristics and well-defined error rates.

Simulation results with in-flight recorded data and Airbus Aircraft Benchmark (AAB, see Chap. 3) are first provided to show the efficiency of the developed technique. Experimental results obtained from the implementation of the developed technique on Airbus Test Facilities (ATF) are also presented. Finally, the technique has been implemented and onboarded on the A380 FCC. The results during more than 70 h of flight confirmed the good robustness the FDD algorithm.

4.2 Industrial State-of-Practice

Consider again the actuator control loop of a moving surface (Fig. 4.2). As also described in Chap. 2, a typical Airbus FCC architecture consists of two separate channels, a command channel (COM) and a monitoring channel (MON). The COM channel provides the main functions allocated to the computer (flight control law computation and the servo-control of moving surfaces). The MON channel ensures (mainly) the permanent monitoring in real time of the COM channel and of all the components of the EFCS (sensors, actuators, other computers, or probes).

The time behavior of runaway and jamming are depicted as in Fig. 4.1. In case of runaway, the control surface position increases until the failure is detected. Then the position remains constant during the reconfiguration time. Reconfiguration means that the surface is controlled by another actuator converted in active mode, and it returns to the controlled position. In case of jamming (the figure on the right), the control surface sticks at its current position when the failure occurs.

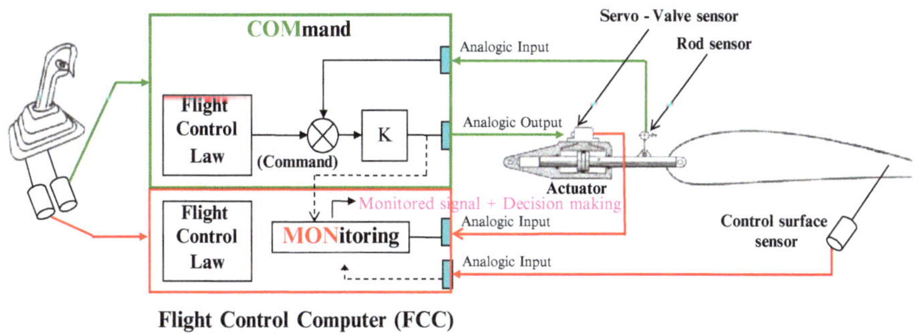

Fig. 4.2 Actuator control loop of a moving surface

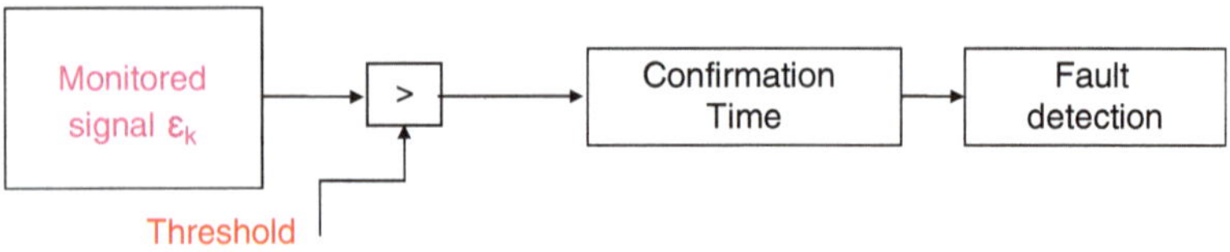

Fig. 4.3 Threshold-based approach for decision making

The current industrial practices for control surface jamming/runaway detection consist mainly in consistency checks between two redundant signals computed in the two FCC channels. If the difference between both signals is greater than a given threshold during a given time, the detection is confirmed. The whole procedure can be again divided in two steps: residual generation and residual evaluation. For runaway case, the residual generation is done by comparing the signal delivered by the servo-valve sensor (Fig. 4.2), which represents an image of the current command sent by the COM channel (see Sect. 3.2) to the actuator, to a kind of theoretical current computed in the MON channel from the actual control surface deflection (generally sensed directly on the control surface by a dedicated sensor) and from the command computed with dedicated redundant sensors in the MON channel (see Chap. 2 and [2, 3]). The error signal ε is computed as follows:

$$\varepsilon = i_{\mathrm{COM}} - i_{\mathrm{MON}} = i_{\mathrm{COM}} - K(u_{\mathrm{MON}} - y_{\mathrm{MON}}) \tag{4.1}$$

where K is the servo-control gain, u_{MON} the command computed in the MON channel and y_{MON} the control surface position acquired in the MON channel (Fig. 4.2); i_{COM} is the command current directly sensed on the servo-valve.

For jamming case, the monitoring signal for fault detection ε, at each sampling time k, is defined according to

$$\varepsilon = |u - y| - |u| \tag{4.2}$$

where y represents the surface position given by the control surface sensor, and u is the command signal provided by the flight control law.

The decision making (Fig. 4.3) for runaway and jamming detection corresponds to a threshold-based approach. Alarms are triggered when the signal resulting from the comparison exceeds a given threshold during a given time window or confirmation time. It can be noted that the pair {threshold/confirmation time} is dissimilar for the two kinds of failure detection.

By setting the threshold, a trade-off must be made between the false alarm and the detection of failures with weak amplitudes. For a small threshold, there is a false alarm risk, and for a big threshold, failures with small amplitudes may go undetected. The above detection techniques ensure the highest level of regulation standard specified by current certification process, provide sufficient fault coverage, and achieve a good robustness without false alarm.

4.3 Need for Improvement

The need for improvement is basically related to the same reasons as discussed in the previous chapter for OFC [4]. Composite materials are used more and more; they involve reduced structural loads on the aircraft. Consequently, the improvement of the current monitoring techniques is a challenging issue, for earlier runaway detection and to decrease the minimal detectable control surface jamming position, while keeping a good level of robustness. For instance, a smaller surface deflection when the runaway is confirmed means less loads generated on the aircraft structure, thus weight saving, better performance, and reduced fuel consumption [5]. From load point of view, aircraft certification is obtained when it is proven that the structure complies with the dedicated regulations. In order to compute the maximum loads to be expected in service, two situations are considered:

- The Time of Occurrence (ToO): moment when the failure occurs; loads must be computed at the time of the failure and immediately after failure. The failure must be detected and passivated quickly; therefore, a small detection time is required by the certification process.
- The Continuation of Flight (CoF): the failure is considered to remain after its occurrence until the end of the flight. The total loads result from the superposition of failure loads and design condition loads (considering maneuvers, gusts, or turbulences). For the aircraft certification, the CoF allows the flight continuation with the detected and passivated failures. Consequently, a long detection time is acceptable, but a small surface position when the failure is confirmed is required.

In order to fulfill the dedicated regulation from certification point of view, an improvement of the current detection techniques is required in order to decrease the detection time (ToO requirement) and the position reached by the control surface when the failure is confirmed (CoF requirement). Clearly, one possible solution for fault detection improvement could be aircraft structure reinforcement, but it is not suitable for being compliant with the aforementioned green objectives. Another solution is to decrease the detection threshold and the detection time, while maintaining a high level of robustness with respect to additional unknown inputs. Note that another important requirement is that the false alarm rate must be extremely low. In fact, when a false alarm is triggered, there is a hand over between the two actuators which control the surface. That means that the healthy actuator is passivated, and then the robustness and the availability of the EFCS are degraded. The non-detection probability must be extremely low since even if the runaway and jamming failure case are very improbable, the consequences are important as explained above. In the following section, it is shown that earlier runaway detection and control surface jamming detection at lowest amplitudes become possible using a simple robust model-based technique. As it has been outlined in Sect. 3.3, from an industrial point of view, use of simple approaches is very important to reduce test phase and also because of certification procedure of aircraft algorithms [6].

Moreover, the procedure developed in the following section provides restricted high-level tuning parameters and low design complexity for easy management of trade-offs and use by non-specialist operators.

4.4 A Dedicated Kalman-Based Solution

4.4.1 *Runaway*

The basic idea is to integrate a dedicated Kalman filter between error signal generation (the residual) and decision making blocks (Fig. 4.4). The monitored signal for runaway case provided by onboard monitoring system ε_k is thus filtered to provide a new monitored signal $\hat{\varepsilon}_k$, where k is the discrete time. The goal of this filter is to early detect abrupt changes while preserving the current robustness performance level.

The following section describes the design procedure. Note that the in-service decision making block (threshold logic and confirmation time) is preserved due to its simplicity and its reliability. Only residual generation is modified.

4.4.1.1 Fault Modeling

A runaway is modeled by the effect that it generates on a control surface. The actuator runaway can be described as an output signal driven directly by the fault. A general description of a runaway behavior $y(t)$ can be written as

$$y(t) = at + b. \tag{4.3}$$

A continuous time state representation of (4.3) has the form

$$\begin{cases} \dot{x}_1(t) = 0 \\ \dot{x}_2(t) = x_1(t) \\ y(t) = x_2(t), \end{cases} \tag{4.4}$$

where x_1 and x_2 are the state variables and the initial conditions given by

$$\begin{cases} x_1(0) = a \\ x_2(0) = b. \end{cases} \tag{4.5}$$

Fig. 4.4 Insertion of a Kalman filter in the monitoring system

Discretization can be done using the Euler approximation and a sampling period T

$$\dot{x}(t) = \frac{x_{k+T} - x_k}{T}. \tag{4.6}$$

If we note the observed output by ε_k, we get thus the discrete-time state model

$$\begin{cases} \begin{bmatrix} x_{1_{k+1}} \\ x_{2_{k+1}} \end{bmatrix} = \begin{bmatrix} 1 & 0 \\ T & 1 \end{bmatrix} \begin{bmatrix} x_{1_k} \\ x_{2_k} \end{bmatrix} + \begin{bmatrix} w_{1_k} \\ w_{2_k} \end{bmatrix} \\ \varepsilon_k = \begin{bmatrix} 0 & 1 \end{bmatrix} \begin{bmatrix} x_{1_k} \\ x_{2_k} \end{bmatrix} + v_k, \end{cases} \tag{4.7}$$

where v_k is the measurement noise and w_k is the process noise. It is assumed that v_k and w_k are both zero mean, stationary white sequences, and Gaussian with covariance matrix

$$\mathbf{E}\left[\begin{pmatrix} w_i \\ v_i \end{pmatrix} \begin{pmatrix} w_k^t & v_k^t \end{pmatrix}\right] = \begin{pmatrix} Q & S \\ S^t & R \end{pmatrix} \delta_{ik}, \tag{4.8}$$

where **E** denotes the expected value operator and δ is the Kronecker delta. It is assumed that $S = 0$ (no correlation between w_k and v_k). Note that the state matrices involved in the model (4.7) are completely known (no model uncertainty). In fact, uncertainties are reported to the unknown inputs.

4.4.1.2 Filter Design

The output ε_k estimation can be realized by a conventional Kalman filter, which provides an efficient recursive computational framework to estimate the state of a process, in a way that minimizes the mean of the squared error. The mechanization equations are well known:

$$\begin{cases} K = P_{k+1/k} C' \left(C P_{k+1/k} C' + R\right)^{-1} \\ \hat{x}_{k+1/k+1} = \hat{x}_{k+1/k} + K\left(\varepsilon_{k+1} - C\hat{x}_{k+1/k}\right) \\ P_{k+1/k+1} = (I - KC)\, P_{k+1/k} \\ \hat{x}_{k+1/k} = A\hat{x}_{k/k} \\ P_{k+1/k} = A P_{k/k} A' + Q \\ \hat{\varepsilon}_k = C\hat{x}_{k+1/k+1}, \end{cases} \tag{4.9}$$

where $P_{k+1/k}$ and $P_{k+1/k+1}$ are predicted and updated estimate covariance matrix, $\hat{x}_{k+1/k}$ and $\hat{x}_{k+1/k+1}$ are predicted and updated state estimate, $A = \begin{bmatrix} 1 & 0 \\ T & 1 \end{bmatrix}$, $C = \begin{bmatrix} 0 & 1 \end{bmatrix}$, and K represents Kalman gain

$$K = \begin{bmatrix} K_1 \\ K_2 \end{bmatrix}. \tag{4.10}$$

In classical Kalman formulation, it is assumed that a complete a priori knowledge of the process and measurement noise statistics (Q and R) is available. Although these characteristics can be inferred from statistical and calibration procedures of the hardware sensing devices, the task is much more difficult for the process noise, since, in essence, it is usually introduced to represent modeling errors. The covariance is usually determined by ad hoc or heuristic approaches, leading to the situations where the filter would not perform in an optimal or desired fashion. Here, Q and R can be considered as design variables to provide a good behavior of the filter, leading to suboptimal Kalman filtering.

From (4.8), the estimated state and output can be rewritten according to

$$\begin{cases} \hat{x}_{k+1} = A\hat{x}_k + K(\varepsilon_k - \hat{\varepsilon}_k) \\ \hat{\varepsilon}_k = C\hat{x}_k, \end{cases} \tag{4.11}$$

where A and C matrices are obtained from (4.7).

Note that the time-varying gain K converges very quickly to a steady-state value. Optimization of the time-varying filter involves optimization of Q and R given a set of specifications. For a steady-state gain, according to (4.10), we get a simple linear time invariant transfer function between the estimated output and the real output

$$\frac{\hat{\varepsilon}(z)}{\varepsilon(z)} = F(z) = C(zI - A + KC)^{-1}K. \tag{4.12}$$

Substituting A, C, and (4.10) into (4.12), we get

$$F(z) = \frac{TK_1 + (z-1)K_2}{(z-1+K_2)(z-1) + K_1 T}. \tag{4.13}$$

Filter (4.13) has low-pass behavior and a steady-state gain equal to 1.

Remark 4.1 As usual, the designer should take care of numerical robustness issues. See, for example [7], for a study on the numerical robustness and performance of existing Kalman filters.

The remaining problem is to determine the optimal filter gain K, which provides a good trade-off between transient response and filtering capacity to ensure runaway detection as fast as possible, while ensuring a good level of robustness. Without adequate tuning of the parameters, the solution is useless. This problem is discussed in the next subsection.

4.4.1.3 Optimization of the Filter Parameters

To find the best tuning of the filter parameters, the problem is formulated as a nonlinear optimization problem under inequality constraints. The stability constraints are derived from the well-known Jury criteria for stability analysis of linear discrete-time systems. The employed resolution methodology consists in a "model matching" approach [8], which seeks to optimize the filter response, based on a desired target model. The methodology is based on the introduction of a reference system response which is considered to be an optimal solution. The theory behind this design procedure can be simply formulated as optimal filter design which optimally approximates the ideal solution provided by the reference model. The optimization problem can be summarized as

$$(\hat{K}_1, \hat{K}_2) = \arg\min_{K_1,K_2} \|M_0 - M(K_1, K_2)\|_l$$

subject to

$$\begin{cases} K_1 > 0 \\ K_1T - 2K_2 + 4 > 0 \\ K_1T - K_2 + 2 > 0 \\ -K_1T + K_2 > 0, \end{cases} \tag{4.14}$$

where M is the output signal of filter (4.13) and M_0 is the reference signal. $l = 1, 2$ represents the norm of the cost function to be minimized. l equal to 2 is used in order to minimize the average behavior between the filtered output and the real output. l equal to 1 is used to minimize sum of absolute differences. The model matching approach can be solved by standard optimization package.

The block diagram for optimization of the filter parameters is given below:

4.4.2 *Jamming*

In this subsection it is shown that the same filter structure (4.13) leads to good results for detection of control surface jamming. However, the filter is not applied on the same residual ε_k, but on the pilot order in order to model the dynamic of the actuator servo-loop (which is given in Sect. 3.2.2.1). The fault indicating signal is now generated according to

$$\varepsilon = |u_{\text{filtered}} - y| - |u_{\text{notfiltered}}| \tag{4.15}$$

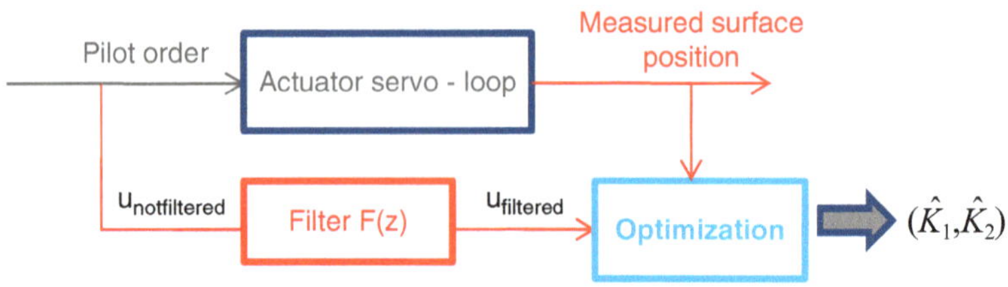

Fig. 4.5 Optimization of detection filter parameters

where $u_{\text{notfiltered}}$ is the command signal provided by the flight control law (pilot order) and u_{filtered} denotes the filtered command signal as shown in the Fig. 4.5. From (4.15), it can be seen that the residual generated for jamming case depends on the difference between the pilot order and the control surface position. By introducing a filter on the pilot order, the difference between the pilot order and the control surface position is minimized, since the filter is tuned to represent the dynamic behavior of the actuator servo-loop.

The optimization of the tuning parameters is again performed within a "model matching" setting by using an appropriate target response. The idea is to optimize filter parameters in such a way that the difference between the output of the filter and the measured surface position is minimized. The optimization problem can be summarized as

$$(\hat{K}_1, \hat{K}_2) = \arg\min_{K_1, K_2} \left\| \bar{M}\big|_{K_a, \Delta P, F_{\text{aero}}} - M(K_1, K_2) \right\|_l$$

where M is the output signal of filter (4.13) and $\bar{M}$ is the servo-controlled hydraulic actuator model resulted from the combination of the following parameters:

- Actuator damping coefficient: $K_a = \left[K_a^{\min} \ \frac{K_a^{\min}}{2} \ \frac{K_a^{\max}}{2} \ K_a^{\max} \right]$.
- Hydraulic pressure: $\Delta P = \left[\Delta P^{\min} \ \frac{\Delta P^{\min}}{2} \ \frac{\Delta P^{\max}}{2} \ \Delta P^{\max} \right]$.
- Aerodynamic forces: $F_{\text{aero}} = \left[F_{\text{aero}}^{\min} \ \frac{F_{\text{aero}}^{\min}}{2} \ \frac{F_{\text{aero}}^{\max}}{2} \ F_{\text{aero}}^{\max} \right]$.

4.5 Experimental Results

4.5.1 Airbus Aircraft Benchmark (AAB) and Real Flight Data

Firstly, simulations are performed using AAB and a real data set recorded during flight tests. The in-flight recorded data comes from an Airbus A380 elevator. An example of real elevator position is given in Fig. 4.6. For confidential reasons, the data are normalized. It can be noted that the control surface position presents a

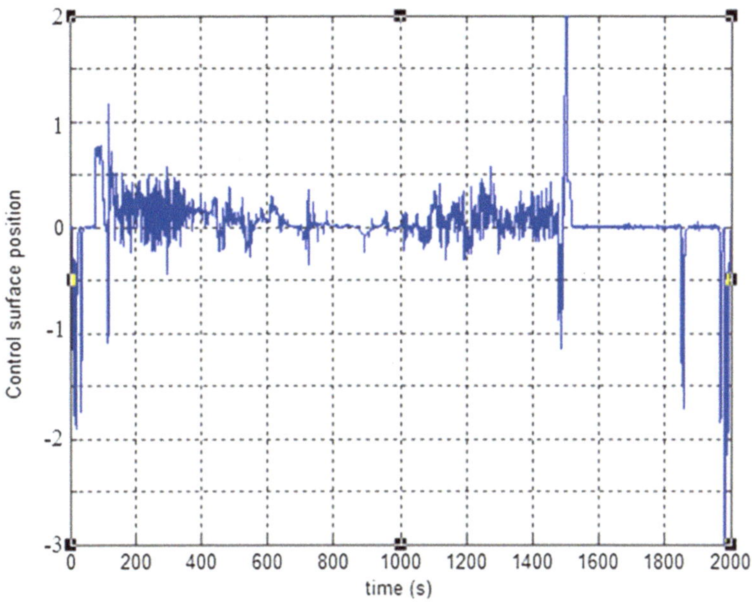

Fig. 4.6 Normalized real surface position

very fast dynamic. Consequently, the simulation results are representative of the method performances during a real flight.

In the remaining of the text, for comparison purpose, two residuals are generated. Residual A is obtained using the current industrial state-of-practice, and residual B is generated using the method described in the previous section, with optimized tuning parameters.

4.5.1.1 Runaway Case

The tuning parameters are optimized by minimizing the difference between the desired response and the real filter response. The real response corresponds to the filter response having the monitored signal as input. The target response is the noise-free filter desired response. Figure 4.7 shows that the reference response is a ramp, which is an image of the control surface position when a runaway occurs. The ramp depends on the runaway dynamic behavior.

Next, fault-free experiments have been run using the AAB. Here, and among a number of available flight scenarios, the focus is put on maneuvers where the aircraft nose may point upward or downward. Figure 4.8 shows the behavior of the residual A and residual B for this flight scenario. Residual peak at $t = 10$ s suggests that aircraft nose points up. As it can be seen, with respect to the residual A, the variability of the residual is reduced, while at the same time, a fast filter response is

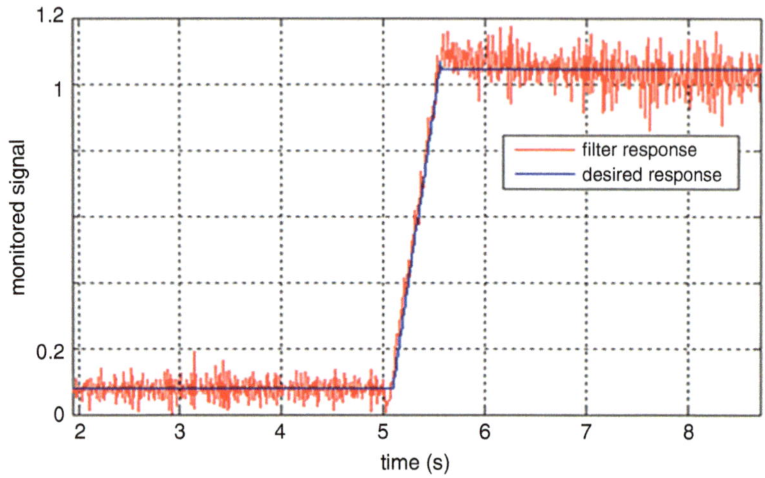

Fig. 4.7 Desired and computed normalized filter responses for a specified runaway speed

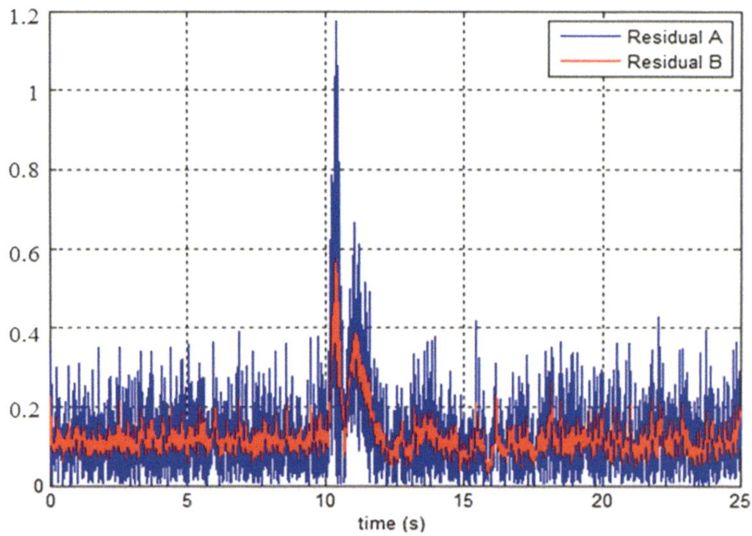

Fig. 4.8 Behavior of the two normalized residuals in fault-free situation using the benchmark

obtained. Thus, the residual B allows a lower threshold without affecting the high level of robustness. This feature suggests that control surface runaway might be detected in an earlier stage.

Fault-free experiments have also been performed using a real data set recorded during flight tests. The same conclusions can be obtained concerning filter performances. Residual B allows a threshold diminution of about 30 % (Fig. 4.9).

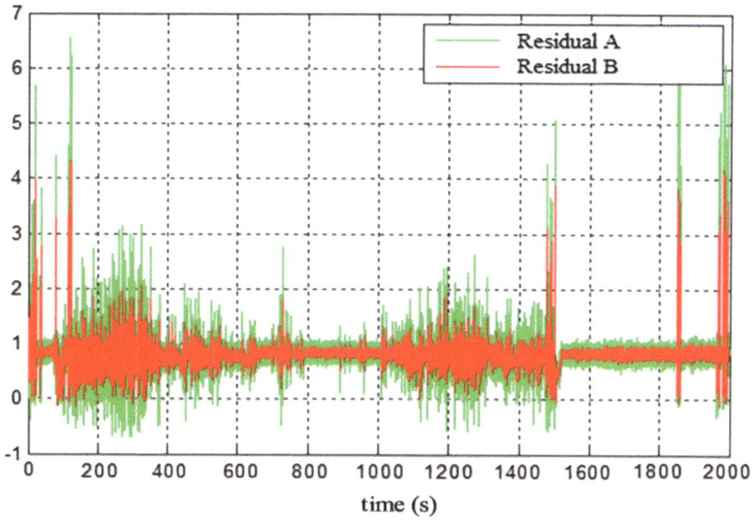

Fig. 4.9 Behavior of the normalized generated residuals using real data sets

Table 4.1 Performance indicators

Runaway speeds (°/s)	Detection delay diminution (%)	Performance indicator improvement (%)
40	≈20	≈25
30	≈23	≈30
20	≈28	≈35
10	≈20	≈25
5	≈28	≈30

4.5.1.2 Performance and Robustness Evaluation

Finally, a number of fault situations are simulated. A performance indicator is calculated as the product between the runaway speed and the difference between failure detection delay before and after filtering. The detection delay is considered to be the difference between the moment when the failure is confirmed and the moment when the failure occurs. This performance evaluation is performed on the real data sets, by simulating various runaway speeds, ranging from 5°/s to 40°/s. The results are summarized in Table 4.1.

The filtering allows a lower threshold without false alarm and a smaller detection delay. The results of Table 4.1 suggest that, by using the Kalman filter, the detection delay is decreased by at least 20 %. Therefore, for all runaway speeds, the performance indicator is positive, showing an improvement or a smaller surface deflection when the failure is confirmed. Table 4.1 illustrates that the surface deflection diminution accounts for about 25–35 % of the maximal surface deflection imposed by the certification process.

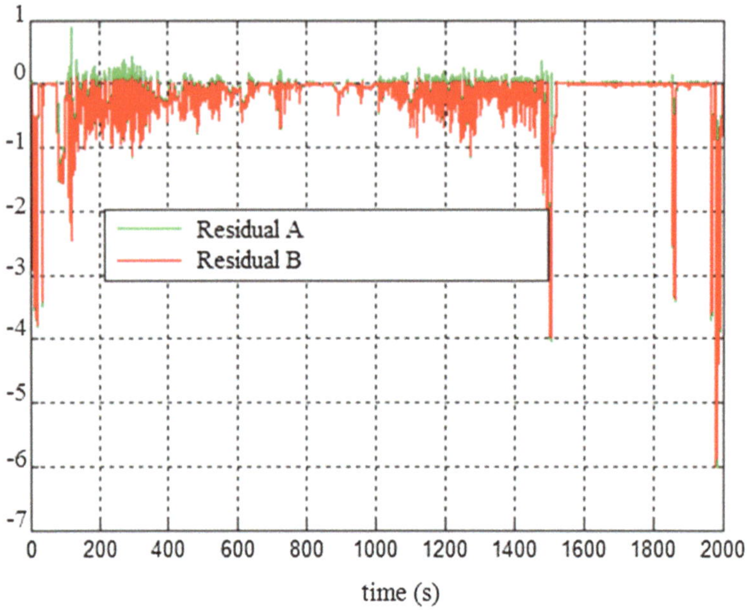

Fig. 4.10 Behavior of the normalized generated residual using real data sets

4.5.1.3 Jamming Case

For the jamming case, using a real data set, the results are depicted in Fig. 4.10. The filter introduced on the pilot order is optimized in order that the difference between the pilot order and the control surface position is minimized. Thus, the maximum value of residual A is reduced, and the control surface stuck at a smaller position can be detected, without degrading the robustness. Residual B allows a threshold diminution with about 50 %.

4.5.2 *Validation and Verification on Airbus Test Facilities*

Significant Verification and Validation (V&V) activities have been performed all along the industrial V-cycle for the technique presented here. Generally, the verification objective is to get assurance that the product (system/equipment) is compliant to its specification. The validation objective is, on the one hand, to obtain the assurance that the specifications are correct and complete and, on the other hand, to get the assurance that the final product is compliant with the customer needs. This section is more dedicated to the verification part. The first phase consists of implementing the new algorithm in the FCC. To accurately specify the functions implemented in the FCC software, a graphical tool is used. A limited set of graphical symbols permits to describe each part of the algorithm in dedicated "functional specification sheets." Then, an automatic generation tool produces the code to be directly implemented in the FCC. Such a tool has as inputs the functional specification sheets and a library of software packages (one package for each symbol used).

The second phase is the integration phase of the V-cycle, where V&V usually proceeds through several steps (see [2] for more details):

- Peer reviews of the specifications and their justification.
- Tests on a desktop simulator using the automatically produced software coupled to an aircraft model.
- Tests on a System Integration Bench (SIB), a test bench used, for example, to tune the servo-control of a given control surface, with simulated inputs and observation of FCC internal variables. This bench also offers the possibility of validating degraded configurations.
- Tests on the "Iron Bird": a test bench that is a kind of very light aircraft, without the fuselage, the structure, but with all system equipments installed and powered as on an aircraft.
- Tests on a flight simulator: a test bench with a real aircraft cockpit, FCC, and coupled to an aircraft model.
- Flight tests on several aircraft fitted with "heavy" flight test instrumentation.

Following encouraging simulation results, the dedicated Kalman filtering with optimized tuning parameters has been implemented in an A380 FCC for integration and flight tests, focusing on an elevator control surface. The algorithm is coded using a restricted symbol library provided by Airbus. The low computational complexity of the given detection method allows for developing a scheme that is only based on basic operators such as delays, multiplications, additions, subtractions, and divisions. Thus, the computational load can be evaluated by using the running time of each symbol. It follows that the developed strategy (the additional Kalman filtering) uses at most 2.2 % of the computing cost allowed for the runaway case, or 0.03 % of the total CPU, and 4.4 % of the computing cost allowed for the jamming case. In the following, the presented results are only dedicated to SIB and flight tests for confidentiality reasons.

4.5.2.1 Experimental Results Provided by the SIB

Elevator runaways have been simulated on Airbus Test Facilities (ATF) with a dynamic ranging from 5°/s to 60°/s. It is confirmed that the Kalman filter improves the state of the art until 40°/s. For higher dynamics, the already in place monitoring ensures better performances. This leads to envisage the possibility of using the two methodologies (with and without Kalman filter) in parallel. As an example, Fig. 4.11 shows the result of a test on Airbus Test Facilities for the jamming case. The objective of this experiment was to determine the minimum threshold without false alarms, in extreme conditions, with maximal pilot order. The filter parameters have been optimized according to the procedure described in Sect. 4.3.2.

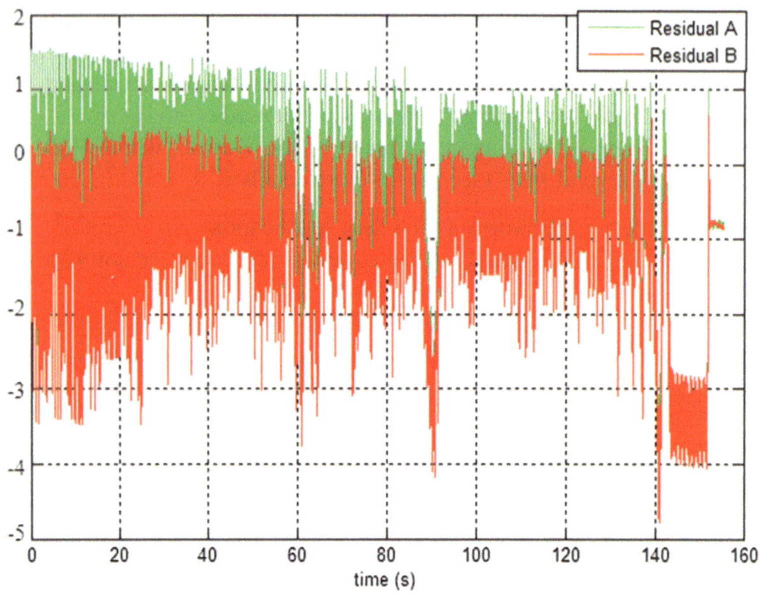

Fig. 4.11 Behavior of the normalized generated residual for jamming detection

4.5.2.2 Real Flight Tests

The level of robustness and performance of the developed method for runaway case are also assessed during real flight tests. Thirty-four flights have been completed, representing more than 70 flight hours. In fault-free situation, and for a specified threshold and confirmation time, the method gives no false alarm. The experiment has been very successful, and no false alarm has been recorded. As an example, the behavior of residuals *A* and *B* can be observed for 1 min of flight (Fig. 4.12).

4.6 Conclusion

In this chapter, we investigated two important failure cases related to aircraft control surfaces. Early and robust detection of such failures are very important from a structural design point of view and also for early system reconfiguration. The focus was to demonstrate that the integration of a simple model-based technique could improve the state-of-practice monitoring performance while maintaining the same level of robustness. Stringent flight requirements are satisfied with low computational cost. With the proposed tuning procedure, designers can use and manage the overall technique quite easily. The developed method has been implemented as a part of the A380 FCC software and provided good results on the Airbus test

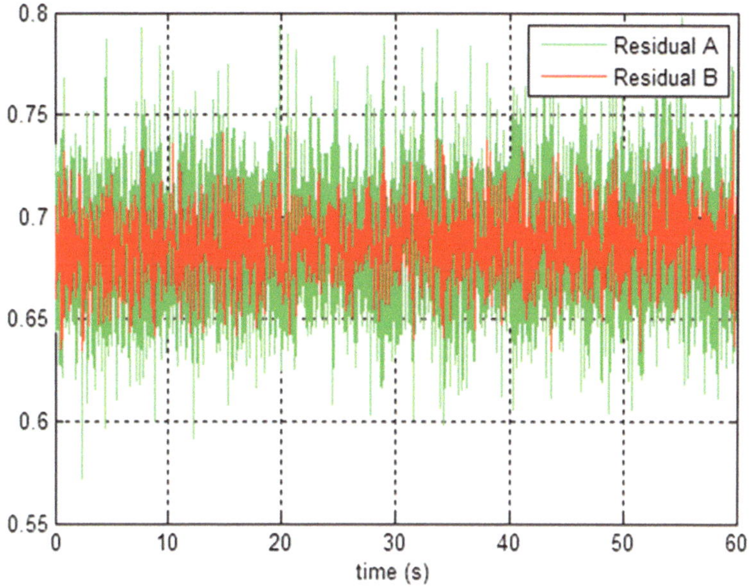

Fig. 4.12 Behavior of the generated residual for runway detection during flight tests

facilities. The robustness of the method was confirmed during real flight tests. The technique seems to be a technologically viable solution for earlier runaway detection and control surface jamming detection at lower amplitude.

References

1. Caglayan AK, Rahnamai K, Allen SM (1988) Detection, identification, and estimation of surface damage/actuator failure for high performance aircraft. In: American control conference, Atlanta, pp 2206–2212
2. Goupil P (2011) AIRBUS state of the art and practices on FDI and FTC in flight control system. Control Eng Pract 19:524–539. doi:10.1016/j.conengprac.2010.12.009
3. Spitzer CR (2001) The avionics handbook. In: Dorf RC (ed) The electrical engineering handbook series. CRC Press, Boca Raton/London/New York
4. Goupil P (2010) Oscillatory failure case detection in the A380 electrical flight control system by analytical redundancy. Control Eng Pract 18(9):1110–1119
5. Gheorghe A, Zolghadri A, Cieslak J, Goupil P, Dayre R, Le Berre H (2013) Toward model-based approaches for fast and robust fault detection in aircraft control surface servo-loop: from theory to application. IEEE Control Syst Mag 33:20–84
6. FAR/CS 25, Airworthiness standards: "Transport category airplane", published by FAA, title 14, part 25, and "Certification Specifications for Large Airplanes", published by EASA, CS-25
7. Verhaegen M, Van Dooren P (1986) Numerical aspects of different implementations. IEEE Trans Autom Control AC-31(10):907–917
8. Zolghadri A, Gheorghe A, Cieslak J, Henry D, Goupil P, Dayre R, Le-Berre H (November 2011) A model-based solution to robust and early detection of control surface runaways. SAE Int J Aerosp 4:1500–1505

Chapter 5
Failure Detection and Compensation for Aircraft Inertial System

5.1 Introduction

This chapter is dedicated to fault detection and isolation of redundant aircraft sensors involved in the computation of flight control laws. The objective is to switch off the erroneous sensor and to compute a so-called consolidated parameter using data from valid sensors, in order to eliminate any anomaly before propagation in the control loop. We will focus on oscillatory failures and present a method for integrity control based on the processing of any flight parameter measurement in the flight control computer (FCC) like, e.g., anemometric and inertial data. One of the main tasks dedicated to the FCC is the flight control laws (FCL) computation which generates a command (position order) to servo-control of each moving surface (see Fig. 5.1). The comparison between the pilot commands (or the piloting objectives) and the aircraft state is used for FCL computation. The aircraft state is measured by a set of sensors delivering, e.g., anemometric and inertial measurements that characterize the aircraft attitude, speed, and altitude. The data is acquired using an acquisition system composed by several dedicated redundant units (usually three). The FCC receives three redundant values of each flight parameter data from the sensors and must compute unique and valid flight parameters required for the FCL computation. This specific data fusion processing, called "consolidation," classically consists of two simultaneous steps (Fig. 5.2): selection or computation of one unique parameter from the three available sources, and, in parallel, monitoring of each of the three independent sources to discard any faulty one. As a consequence, the consolidation allows reliable flight parameters computation with the required accuracy by discarding any involved failed source.

Current consolidation state-of-practice allows the designers to be compliant with stringent specified regulations. However, for structural design optimization and for easier-to-handle future aircraft programs, it could be required to prevent the propagation of spurious signals to the control surfaces, even of small amplitudes, and to extent the measurement availability. In aeronautical engineering, the fault-hiding paradigm [1], also known as virtual sensor paradigm, is used to achieve

A. Zolghadri et al., *Fault Diagnosis and Fault-Tolerant Control and Guidance for Aerospace Vehicles: From Theory to Application*, Advances in Industrial Control,
DOI 10.1007/978-1-4471-5313-9_5,

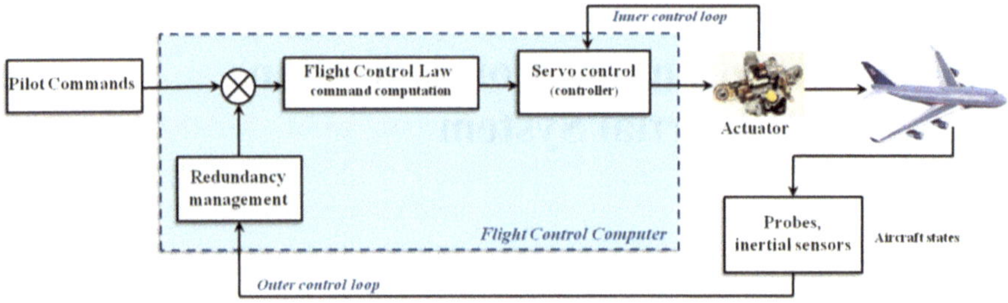

Fig. 5.1 Flight control law computation

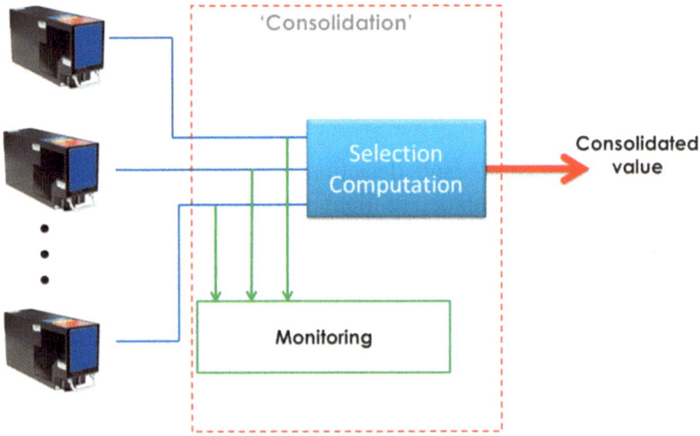

Fig. 5.2 General principle of the consolidation process

a fault-tolerant process. The idea is to put a consolidation (or reconfiguration) block (redundancy management block in Fig. 5.1) between the faulty plant and control laws to hide the faults from control laws. A fault-tolerant management system can be used to check the consistency of all sensor outputs to diagnose a failed source and to inhibit its effect, typically by using a majority-voting or a weighted mean method [2] or soft-computing approaches [3]. The main advantages of this architecture are that the design and integration are relatively simple while providing acceptably efficient detection of system failures [4]. Today, the well-known majority-vote-based techniques are the standard industrial practice and used in current aircraft certification process. Three sources are usually used for the majority vote. If a source fails, then its contribution has to be removed from the consolidation. The fault tolerance is however not guaranteed with only two valid sources. Moreover, if two sources become faulty simultaneously (i.e., really at the same sample time), the consolidation could be tricky to achieve. Some advanced data fusion techniques [5–8] can be used to overcome the issue, but they are usually computation hungry that limits their on-board application. The increasing number and power of embedded calculators and decentralized design strategies for upcoming programs will allow, in a near future, the implementation of more complex processing techniques.

Usual failures of the flight parameters sensing system include (but are not limited to) oscillations, bias, freezing, drift, loss of accuracy, and calibrations errors (scaling) [9]. In this chapter, the focus is put on the additive oscillations appearing on output signals. These failures are referred as Combined Oscillatory Failure Cases (COFC) in contrast of Oscillatory Failure Cases (OFC) that impact only one control surface (see Chap. 3 and [10–12]). COFC can possibly impact several control surfaces. The measurements provided by the corrupted source could propagate downstream the control loop computation and may cause under some circumstances unwanted oscillations of the control surfaces. That is why OFC and COFC (of unknown amplitude and unknown frequency) must be detected as quickly as possible, usually in less than a limited number of periods of oscillation (see Chap. 3).

In the literature, oscillation detection is usually handled using time-based or statistical methods. In [13], a real-time detection method based on controller error zero crossing was presented. In [14], the same idea was extended to off-line analysis. A statistical approach was proposed in [15], based on the autocorrelation function, since the autocorrelation of a periodic signal is also periodic. In [16], the zero-crossing idea was applied to filtered auto-covariance data to check presence of oscillations in selected frequency ranges. A review of indirect approaches was proposed in the following [17]: closed-loop performance assessment is used to detect abnormal behavior, and oscillation isolation is performed on suspected loops [18]. A similar approach, based on higher-order statistics and visual representation is presented in [19]. Some sinusoidal component's estimation methods rely on the computation of some form of a discrete Fourier transform, as proposed in [20–22]. Other authors use principal component analysis-based approaches: in [23] authors use Karhunen–Loeve basis, and in [24] independent component analysis is used to perform oscillation detection on multiple signals. In [25], oscillation detection is performed on robot structures using fast Fourier transform.

Majority-voting-based techniques are not commonly used for OFC detection. However, majority-voting-based techniques can detect any type of failure with three valid sensors, not only OFC failures, but also freezing or drift errors. The sensor management system proposed in this chapter aims at the improvement of the existing three sensor acquisition systems (also referred as triplex). In particular, the system is able to detect COFC and to compute a non-corrupted parameter with only two valid sensors. The proposed solution is based on a hierarchical FDI structure which takes in consideration the number of healthy sensors in the system:

- When more than two sensors are available, soft-computing techniques are used for fault detection and for data consolidation. An appropriate criterion is proposed to ensure proper OFC isolation.
- When there is no possible majority or when COFCs occur on two sources, harmonic filter-based FDI process detects failures node by node.

The novelty in the proposed method is the use of a particular OFC indicator based on the harmonic filter developed in [26]. The idea is to transform the fault signature from sinusoidal to steplike, which greatly improves the detection rate for low amplitude oscillatory failures. A study of different abrupt change detection techniques is carried out. The obtained results are analyzed in terms of measurement noise cancellation and avoidance of false alarms. The benefit of the presented method is to improve the consolidation process with a fault detection and isolation approach when only few sources (less than three) are valid. The approach is validated on a real recorded flight data set.

The remainder of the chapter is organized as follows: the next section reviews briefly the current industrial practice. Section 5.3 presents the proposed structure based on a mixed data fusion/fault detection and isolation method for consolidation. The soft monitor and the harmonic filter are described. Section 5.4 presents the simulation results obtained using a normalized real flight data set. The material is mainly underpinned by the published paper [12].

5.2 Failure Detection and Isolation in Aircraft Inertial System

5.2.1 *Problem Statement*

Given three flight parameter aircraft sensors, and considering possible sensor failures, the objective of this chapter is to improve the classical majority-vote triplex monitoring structure in situations where less than three valid sensors are available. The focus is put on COFC since the other sensor errors are covered in the literature [3] for the case of three sources. In the following, three failure cases are investigated:

- Loss of one source with three valid sources available initially (where the classical majority-based approach is appropriate)
- Three valid sources initially and two sensors affected simultaneously by the same COFC
- Loss of one source with two valid sources and a corrupted source initially (where the majority-based approach is impossible)

5.2.2 *Classical Triplex Monitoring*

Current industrial practices involve triplex voting schemes as in [27, 28]. The very basic principle is to choose the median value as the voted value (i.e., to give a null weight to the extreme values and a weight equal to one to the median value). Another common principle is to sort output signals and to give a 0.5 weighting to the source

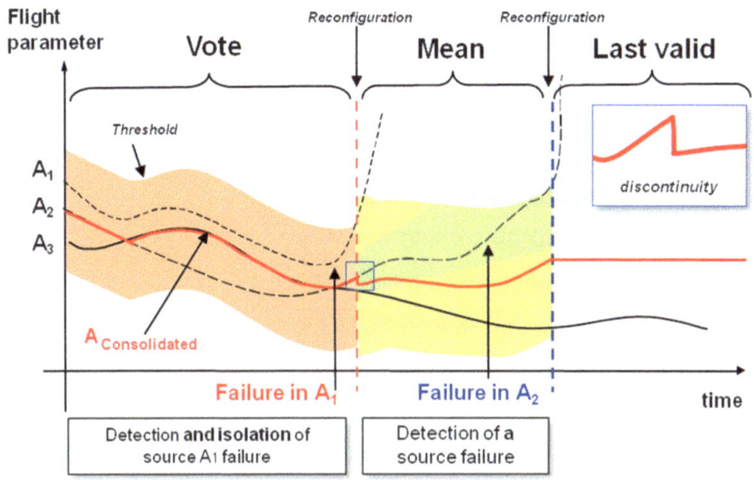

Fig. 5.3 General principle of triplex monitoring

providing the second (median) value of the parameter and a 0.25 weighting to the two other sources, then adding the results to obtain the value of the parameter. A threshold, centered on the obtained value, is used to detect the occurrence of a failure. When a COFC occurs, the corrupted source is detected when the provided measurement stays outside the threshold for a specific amount of time.

If only two sources are valid, the consolidation is generally performed by choosing the mean value (or a weighted average) of the two measures. Notice the discontinuity when switching from vote-based consolidation to the mean-based one. The monitoring is performed by comparing each remaining source with the mean. If the difference between the two signals is superior to a specified threshold during a given time, then, as it is impossible to identify which source is the healthy one, the two sources are eliminated. In that case, several solutions are possible: either to consider the parameter of interest as fully lost or to keep the last correct value. Fault isolation is possible only when the three sources are valid, and all measurement updates are lost when two sources are corrupted, since it cannot be decided which source can be trusted. Also, thresholds must be chosen off-line for all the possible flight scenarios, involving long and costly experiments. As it can be seen in Fig. 5.3, transients appear on the consolidated parameter.

5.3 Enhanced Detection and Compensation Scheme

In order to improve the classical triplex monitoring described above, a mixed data fusion (DF)/fault detection and isolation (FDI) method is proposed. The overall structure of the proposed method is shown in Fig. 5.4. There are two major

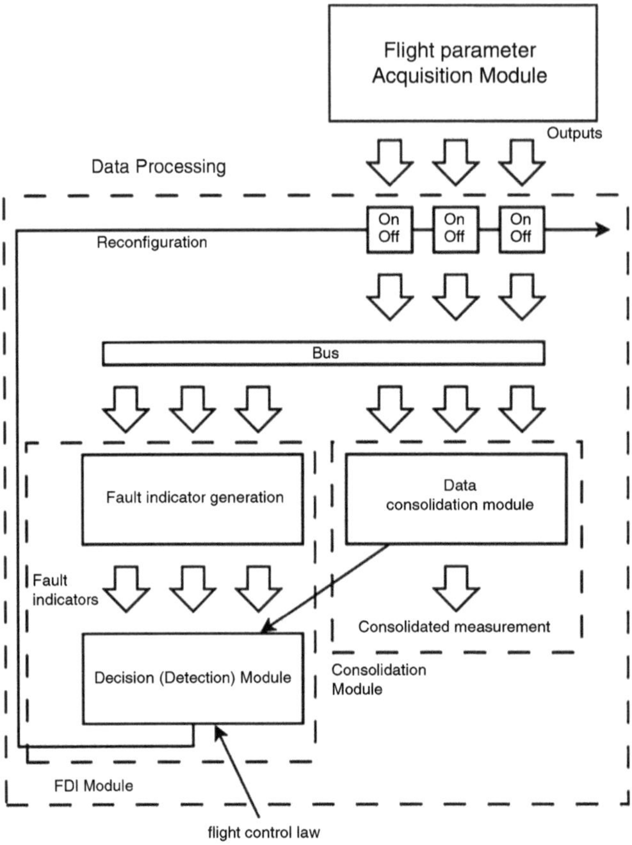

Fig. 5.4 Overall structure of proposed methodology

components: the fault detection and isolation component including a filter and a decision-making module and a second component which is the consolidation module. The role of the FDI module is to detect the source corrupted by the OFC and to switch it off as soon as possible, while the consolidation module gives an accurate estimation of the parameter based on the measurement data provided by the valid sources. The knowledge of the flight control law is used to discriminate between the COFC and a possible (normal) sinusoidal control law input of the pilot. The consolidation module also provides FDI functionality in the three valid sources scenario, detecting and isolating single OFC, freezing and drift failures.

There are two options to manage interaction of modules: if the error on the consolidated parameter is to be minimized, and enough computational power is available, then the two modules can be executed in parallel, since COFC occurrence is possible. Alternatively, in order to minimize the computation time and if an increased isolation delay and a less good parameter consolidation are acceptable for a limited period, the FDI module is switched on only when a detection occurs, since the consolidation module provides sufficiently efficient fault detection with three valid sources initially. After the occurrence of an OFC, the consolidation module is switched off, and the consolidated parameter is computed using an average mean of all valid sources.

5.3.1 OFC Detection and Isolation Module

As already mentioned, a number of methods exist for oscillatory fault detection. In the open literature, time-based and statistical approaches are the two main categories. Some of the most popular methods are detailed and compared below (see also [29]):

- *PCA Based* [23, 24]: The Karhunen–Loeve [23] based approach provides online detection of oscillations. It is robust to time window length, but needs to set a detection threshold, which means noisy data may be difficult to use. Also the detection during transients is not reliable which will cause false alarms. The second reference is an off-line approach based on independent component analysis. It performs well with noisy data but transients are a problem.
- *Zero Crossing and Variants* [13, 16]: Data zero crossing-based approaches are independent from threshold, which makes them suitable for noisy environment. The detection can catch regular oscillations in steady-state data. The detection is carried out with a delay of a few oscillation half-periods. The method is strongly dependent on window length. A possible variant is to use data auto-covariance to improve oscillation detection, at the expense of going off-line.
- *DFT Based* [20–22, 25]: Fourier transform-based methods ignore transients (no detection during transients) and do not require threshold tuning or model knowledge. The method is very dependent from the window length and requires a relatively large CPU time.
- *Autocorrelation Function* [15]: Autocorrelation-based oscillation detection is an off-line approach, well suited to noisy data. However, it is more suited to detect larger oscillations and performs poorly when oscillations are small.
- *Indirect Approaches* [17]: These methods are off-line methods. Performance assessment, for instance, makes possible the detection of an abnormal loop, and an additional isolation step is needed to detect an oscillation. These methods are robust, but with the additional oscillation isolation, they can become computationally hungry.

For the considered application here, off-line analysis is not relevant, so only online methods will be considered. Transients are often problematic, so if the model is unknown, the best choice is to use DFT-based, Kahrunen–Loeve, or zero-crossing methods (if the detection delay is acceptable). The DFT and the PCA approaches require important CPU time though, which can be problematic for on-board implementation, today, flight computers can assign only a few percents of their time for OFC detection.

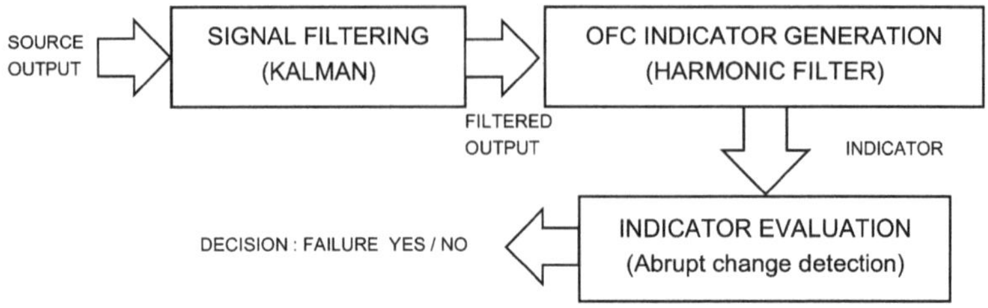

Fig. 5.5 FDI system

In this chapter, the proposed dedicated OFC detection method relies on a particular characteristic of the harmonic filter proposed in [26]. For this problem, the parameter estimation functionality of this filter is useless, especially considering the important convergence time and the additional computations. However, the filter is very sensitive to new harmonics appearing in input signal spectrum. This functionality is used to design a selective OFC indicator. The measured signals are noisy, so the direct fault indicator generation is difficult. An additional filtering component is considered in the form of a near optimal steady-state filter similar to the filter developed in [30, 31]. Finally, an abrupt change detection method is used to perform COFC detection. Three methods are presented and compared: a simple gradient algorithm, a robust differentiator, and a slope change detection procedure. Notice that the thresholds are computed off-line, in fault-free conditions. The FDI mechanism is represented in Fig. 5.5.

5.3.1.1 Selective Filter for OFC

Consider the following fault signal:

$$d(t) = a\cos(\omega_0 t + \phi) \tag{5.1}$$

where a, ω_0, and ϕ are, respectively, the amplitude, frequency, and phase of the sinusoidal signal $d(t)$. The corresponding state-space model is

$$\begin{cases} \begin{bmatrix} \dot{x}_1 \\ \dot{x}_2 \end{bmatrix} = \underbrace{\begin{bmatrix} 0 & 1 \\ -\omega_0{}^2 & 0 \end{bmatrix}}_{A} \begin{bmatrix} x_1 \\ x_2 \end{bmatrix} \\ d = \underbrace{\begin{bmatrix} a & 0 \end{bmatrix}}_{C} x + v \end{cases} \tag{5.2}$$

where v is a Gaussian noise. The Kalman filter for this system is described as

$$\dot{\hat{x}} = A\hat{x} + H(y - C\hat{x}).$$

In [30], an appropriate value of H for this model is given by

$$H = 2\xi\omega_0 \begin{bmatrix} 1 \\ 0 \end{bmatrix}.$$

It is a reasonable choice leading to a suboptimal steady-state filter, with ξ being a small constant. The complete equation of this filter is given by

$$\dot{\hat{x}} = \begin{bmatrix} -2a\xi\omega_0 & 1 \\ -\omega_0{}^2 & 0 \end{bmatrix} \hat{x} + 2\xi\omega_0 \begin{bmatrix} 1 \\ 0 \end{bmatrix} y. \tag{5.3}$$

The filter is a stable second order between the output and $\hat{x}_1$

$$G(s) = \frac{\hat{X}_1}{Y} = \frac{2\xi\omega_0 s}{s^2 + 2a\xi\omega_0 s + \omega_0{}^2}.$$

This simple system provides near optimal performance when filtering sinusoidal signals around ω_0 frequency. The selectivity of the filter is tuned through the parameter ξ. In the following, the parameter a is set to $a = 1$, since the scaling will not hinder the filter's performance (see [30], pp. 289–290).

5.3.1.2 OFC Indicator Generation

A classical decision-making method in FDI is to check threshold crossing by the fault indicator signal. The usual problem here is to determine appropriate threshold values in order to simultaneously maximize fault detection ratio and to minimize detection delay and false alarm rate. This problem has many possible solutions for fault signatures that induce abrupt changes in the system behavior; check, for instance, [32, 33]. But for oscillatory failures the problem is much more difficult, as classical abrupt change detection methods are ill-suited for detecting smooth changes in the system signals, especially in case of low amplitude oscillatory failures and noisy measurement. A possible solution to this problem is to transform the sinusoidal influence of the COFC on the system into a steplike influence. The determination of the detection threshold is simplified and is solely based on the parameters of the measurement noise, usually available and depending on performance of sensors. This transformation is obtained using an appropriate additional filter. From this perspective, the work reported in [26] will constitute the basis of the OFC indicator. In the original paper, an estimator is proposed to provide accurate estimation for a sinusoidal component in a noisy signal. It is shown that the outputs of the estimator converge asymptotically to the correct values in a given time. The core of the estimator is a third-order nonlinear filter described by the relation

$$\begin{cases} \dot{x}_1 = Kx_3\left(-2\alpha\left[-2\alpha x_3 - \alpha^2 x_2 + u\right] - \alpha^2 x_3\right) - Kx_3{}^2 y - K\left[-2\alpha x_3 - \alpha^2 x_2 + u\right]u \\ \dot{x}_2 = x_3 \\ \dot{x}_3 = -2\alpha x_3 - \alpha^2 x_2 + u \\ y = x_1 + Kx_3 u \end{cases} \tag{5.4}$$

The coefficients K and α are tuning parameters, with K acting like a gain and α acting like a damping ratio. When $t \to \infty$, the relation $\omega = \sqrt{|y|}$ is an accurate estimation of the pulsation of an input oscillatory signal from the form (5.1). The remaining parameters (amplitude and phase) are given by

$$a = \sqrt{\frac{\hat{\dot{u}}^2}{y} + \beta^2}$$

$$\phi = y - \beta$$

with

$$\beta = \frac{\hat{\ddot{u}}}{y}$$

and

$$\hat{\dot{u}} = yx_3 + 2\alpha\left[-2\alpha x_3 - \alpha^2 x_2 + u\right] + \alpha^2 x_3$$

$$\hat{\ddot{u}} = y\left[-2\alpha x_3 - \alpha^2 x_2 + u\right] + 2\alpha y x_3 + \alpha^2\left[-2\alpha x_3 - \alpha^2 x_2 + u\right]$$

where $\hat{\dot{u}}$, $\hat{\ddot{u}}$ are the expressions of the input derivative estimates. The OFC indicator proposed in this work is based on the nonlinear filter described in Eq. (5.4). The main difference is that the indicator is based only on the most OFC sensitive parameter, noted y in the original paper, dismissing the parameters used for estimation in [26]. The filter is sensitive to the input of an oscillatory signal and reacts by a slope change of the output y. The proof is omitted for the sake of brevity (see [26]). The complete equation of the OFC indicator is given by

$$\begin{cases} \dot{x}_1 = Kx_3\left(-2\alpha\left[-2\alpha x_3 - \alpha^2 x_2 + x_4\right] - \alpha^2 x_3\right) - Kx_3{}^2 y - K\left[-2\alpha x_3 - \alpha^2 x_2 + x_4\right]x_4 \\ \dot{x}_2 = x_3 \\ \dot{x}_3 = -2\alpha x_3 - \alpha^2 x_2 + x_4 \\ \dot{x}_4 = -2\xi\omega_0 x_4 + x_5 + 2\xi\omega_0 u \\ \dot{x}_5 = -\omega_0{}^2 x_4 \\ y = x_1 + Kx_3 x_4 \end{cases} \tag{5.5}$$

Remark 5.1 The filter presented in [26] was proposed primarily for identification purposes. An interesting enhancement would be to optimize the discussed filter with fault detection objectives in mind, maximizing the sensitivity, for example.

5.3.1.3 Abrupt Change Detection

Abrupt change detection is a decision-making process which is usually based on threshold logic of a decision function [34]. If there are no uncompensated unknown effects on the residuals (perfect case, very low noise), then the thresholds diminish to zero. Otherwise, thresholds different from zero must be assigned. In this case, robust residual evaluation is the only way to keep the false alarm rate small with an acceptable sensitivity to faults (see Chap. 2). Classically, abrupt change detection can be accomplished in many ways, for example, by statistical data processing, data reconciliation, correlation, pattern recognition, fuzzy logic, or adaptive thresholds. In this case, the harmonic filter provides a good discrimination for all non-periodic outputs, and behaves in an easily predictable way when a periodic signal is present in the processed data. Using the OFC indicator provided by the relation (5.5), or its energy, the detection is possible using any abrupt change detecting approach; check, for example, [35] for generic problems. A classical gradient algorithm on the OFC indicator, coupled with a threshold-crossing detection, is a possible choice and gives satisfactory FDI performance with a low computational cost. However, for very low-frequency OFC failures, any measurement noise makes the detection difficult. Two different approaches, with reasonable computational costs, are compared to circumvent this issue: slope change detection, similar to the method used in [36] and the discrete-time robust derivative estimator developed in [37]. These two techniques were selected because of the reasonable performance/complexity balance.

Robust Derivative Estimator

The derivative estimator used is a discrete-time high-gain observer (see, for example, [37]). The discrete transfer function is given as

$$T(z) = \frac{2\beta}{2z.epsi; + \beta \Delta T} \times \frac{z-1}{z + \frac{1 - \frac{2z.epsi;}{\beta \Delta T}}{1 + \frac{2z.epsi;}{\beta \Delta T}}}$$

where ΔT is the sampling time and β, *z.epsi;* are tuning parameters. The behavior of this filter is fixed by the ratio $\Delta T/z.epsi;$. High values (superior to 1) are taken when the noise level is low and low values when high-level noise is present.

Slope Change Detection

The method is straightforward. Assuming that on a sufficiently short period, the evolution of the residual is linear; the data set of interest is decomposed in fixed time windows. A linear regression is performed to approximate the data in the window. At the end of each window, the slope value is compared to the previous estimation. If the difference is superior to a predetermined threshold, then OFC detection is triggered.

Online implementation is straightforward. The data is acquired on a time window. A linear regression is performed and the obtained slope is compared to the previous estimation. If the difference is not noticeable, OFC detection will not trigger and the new estimation is stored in memory. The benefit of this approach is a good robustness to the measurement noise, and the principal drawback is that the detection delay will be at least the size of the time window.

5.3.2 *Consolidation Module*

The consolidation module is similar to the system proposed in [3] with a specific OFC isolation component. The fusion between different sources is performed using a fuzzy logic approach called soft voting. To each source is assigned a weight corresponding to the amount of trust it is credited, and the consolidated signal is the weighted average of all valid sources (Eq. 5.6). This approach provides substantial benefits compared with other majority-voting approaches for a reasonable computational over-cost. The principal benefit of soft voting is the fact that the corrupted source is suppressed before the detection. Hence, there is no discontinuity on the consolidated parameter when the detection occurs and the corrupted source is switched off [38]. It is a soft fault compensation where the consolidated measurement is given by

$$S_{\text{vote}} = \sum_{i=1}^{n_{\text{valid}}} w_i S_i \tag{5.6}$$

with w_i representing the weight to the source S_i and n_{valid} corresponding to the number of valid sources. The weight w_i is computed from the membership degree $\mu_i \in [0, 1]$ assigned to each measurement

$$w_i = \frac{\mu_i}{\sum_{j=1}^{n_{\text{valid}}} \mu_j}. \tag{5.7}$$

The computation of μ_i is shown in Fig. 5.6. Each membership function (in blue) is centered on the value provided by the corresponding source. Thus, to compute μ_1 the function is centered around the value provided by the first source q_1, and the

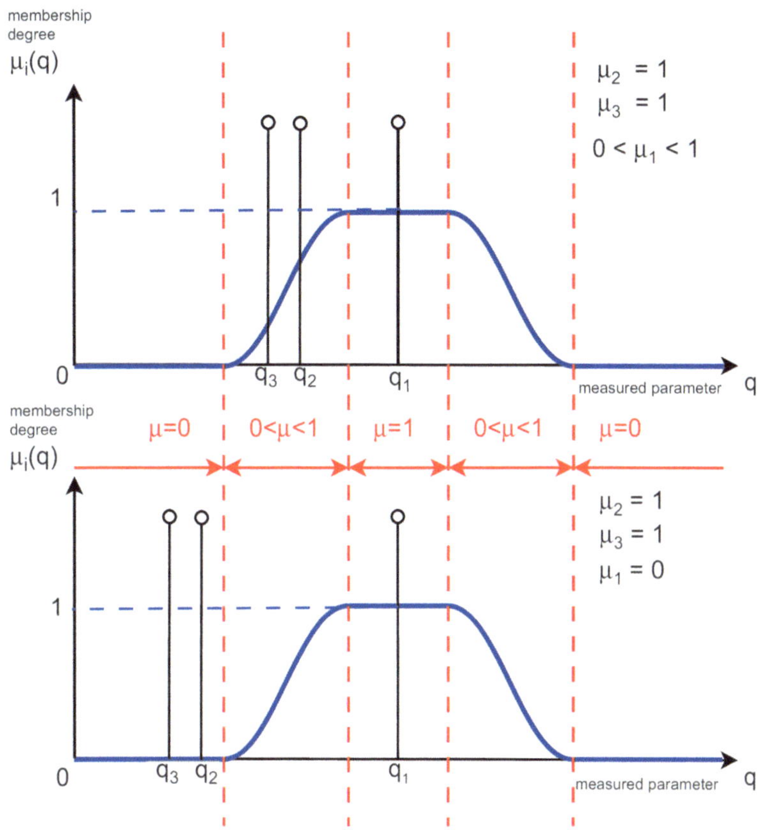

Fig. 5.6 Membership degree computation

membership degree of the source is given by the largest membership degree of the remaining valid signals. For example, in Fig. 5.6, the first case shows that $\mu_2 = 1$, since the largest membership degree of sources 1 and 3 is 1; indeed q_2 and q_3 are very close. The second case shows that $\mu_1 = 0$ because q_1 is far enough from q_2 and q_3. In both cases, only the closest q_i matters.

$$\mu_i = \max_{i \neq j} \left(\mu_i(q_j) \right) . \tag{5.8}$$

The majority-voting concept is used in soft voting as it is used in conventional consolidation. The difference is the contribution of the measurement signals: in the conventional scheme, the contribution of the faulty signal is limited while in soft-voting scheme it is reduced. The direct consequence of the "limitation" versus the "reduction" is the discontinuity of the consolidated measurement appearing when a failed source is switched off in the classical voting scheme. By using such scheme, soft fault compensation can be obtained.

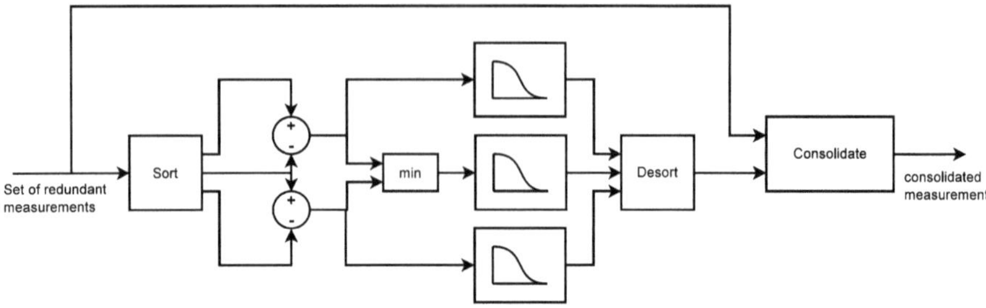

Fig. 5.7 Soft voter

Table 5.1 Soft-monitoring procedure

if $\mu_i = 1$	then	$\text{count}_i = \text{count}_i - 1$
if $0 < \mu_i < 1$	then	$\text{count}_i = \text{count}_i$
if $\mu_i = 0$	then	$\text{count}_i = \text{count}_i + 1$

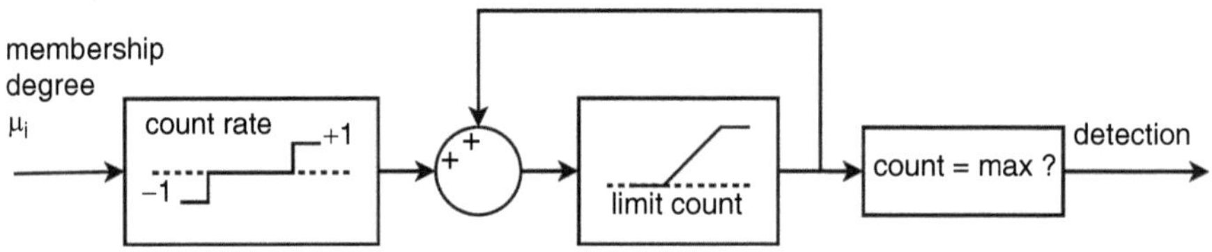

Fig. 5.8 Soft monitor

In the soft-voting-based consolidation, the consolidated measurement remains smooth in all cases. The overall structure of the soft-voting block is shown in Fig. 5.7. The measurements are sorted by value, from the smallest to the greatest (the "sort" function). Then, subtractions are carried out, and the results are weighted using membership functions in order to compute membership degrees. The "desort" function arranges the resulting membership degrees to match the order of the initial measurements vector. The consolidation is a straightforward multiplication of the vectors element by element and the sum of the result, which actually gives the consolidated flight parameter.

The monitoring component is based on a counter associated to each source (Table 5.1) and threshold-crossing detection logic. A threshold is set to the maximal admissible consolidation error to detect any sensor divergence including freezing and drift failures. The overall scheme of the monitoring block is shown in Fig. 5.8.

The counter is not a function of the difference between the consolidated value and the ith measurement as in the conventional scheme, but it is a function of the difference between the measurements, which are expressed by the membership degrees of each source. Therefore, no transients occur when a source is switched off since its contribution to the consolidated parameter was already nil. It is a great advantage from fault compensation point of view.

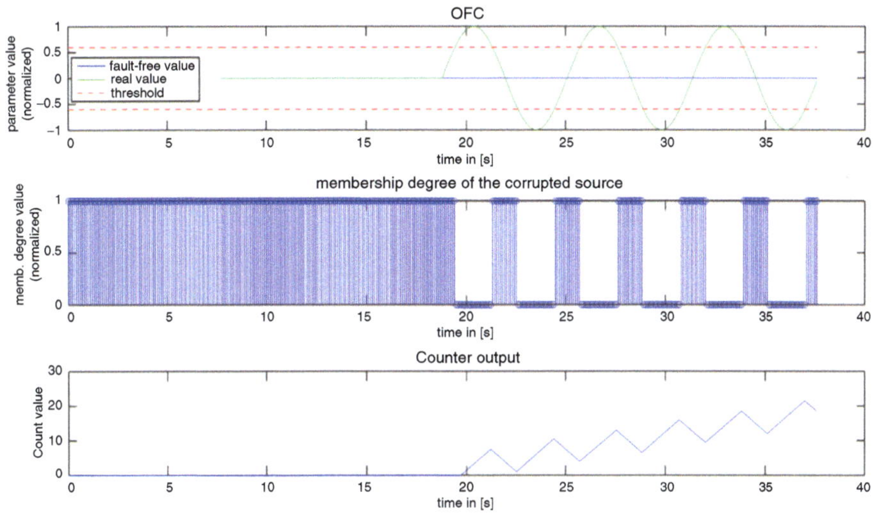

Fig. 5.9 OFC detection by the soft monitor

When an OFC occurs, the detection is triggered quickly using threshold-crossing detection logic, but the oscillation is not identified yet. Since the failure is periodic, the isolation is performed by monitoring the periods between transitions from 1 to 0 (or 0 to 1) of a membership degree μ_i as shown in Fig. 5.9. It is a variation of the zero-crossing oscillation detection. When four successive "1 to 0" transitions of μ_i show periodicity, an OFC is confirmed. The same result can be achieved by tracking periodic increase and decrease of the count. The isolation delay may be important, yet, the effect of the failure is naturally accommodated by the soft-voting approach [3] reducing the impact of the delay for a single OFC. For COFCs, while the detection is possible, fault isolation should rely on dedicated FDI module.

Remark 5.2 In case of a drift or a freezing failure, the counter evolves as shown in Fig. 5.10. The dashed lines represent detection thresholds. The figure shows that the soft monitoring can successfully detect this type of failures, when the failure occurs on a single source. Multiple sensor drifts and freezing are not covered though. The soft monitor will detect the failures, but will not correctly identify the faulty sources (the reason is explained in the discussion on COFC illustration). A possible solution to this problem is to add a specific FDI module, tuned to detect drifts or freezing. Freezing can also be detected by checking just the derivative.

Remark 5.3 In the classical majority-voting approaches, the counter will detect large amplitude OFC, but the presence of OFC cannot be confirmed. The detection is based on threshold-crossing detection, so a drift will be detected in the same way.

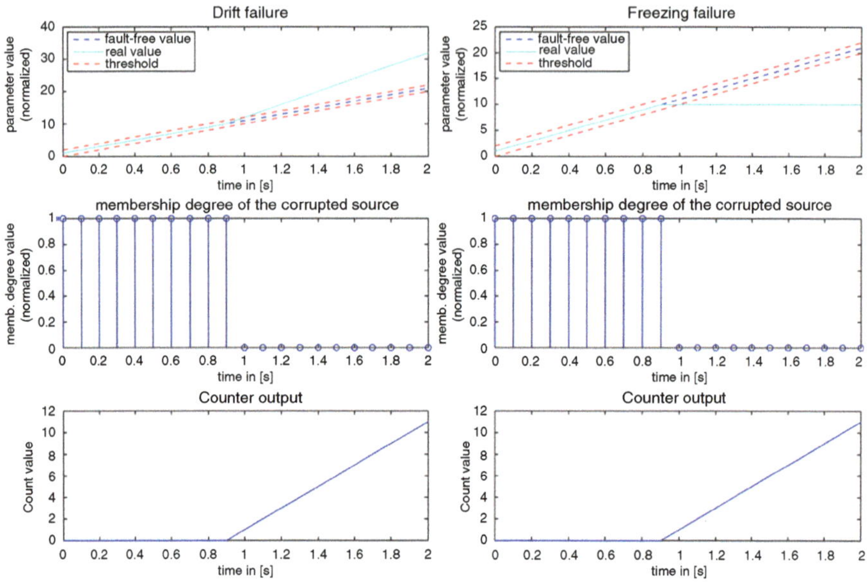

Fig. 5.10 Drift and freezing detection by the soft monitor

5.4 Simulation and Experimental Results

The simulations are firstly performed using a commercial aircraft benchmark provided by Airbus during SIRASAS[1] project (see Chap. 1). The results have been judged satisfactory and encouraging. Next, to assess its benefit in realistic flight environment, the proposed monitoring scheme was applied to a real recorded flight data set. In the following subsections, only the results obtained by using the flight data set will be presented.

5.4.1 *Simulation Setup*

The processed data set is recorded during a real flight test of a large civil commercial aircraft. This flight test is used to show the good performance of the FD/FDI scheme for both lateral and longitudinal modes. For industrial reasons, all simulation parameters and results are normalized. The measured parameter is the vertical load factor (inertial measurement).

[1] Innovative and robust strategies for spacecraft autonomy.

Table 5.2 Influence of K and α selection on FDI performance

K/α (diff. ratios)	100/10	50/10	20/10
Av. detection delay (s)	1.25	1.02	0.97
K/α (diff. ratios)	10/10	10/20	10/50
Av. detection delay (s)	50 % ND	50 % ND	100 % ND
K/α (diff. ratios)	100/50	40/20	20/10
Av. detection delay (s)	100 % ND	66 % ND	0.97
K/α (diff. ratios)	10/5	2/1	
Av. detection delay (s)	4	100 % ND	

ND no detection

Two cases are considered:

- A single failure occurs on the first inertial sensor with all units initially being healthy. The detection and the isolation of the OFC are performed using the soft-monitoring module.
- Simultaneous double failure on the first and second inertial sensors. The detection and the isolation of the COFC are based on the analysis of the residuals corresponding to each source.

Remark 5.4 The second case, where only one valid source is still available, is very similar to the case when two successive OFCs occur on the first and then the second sources, since the FDI method proposed here does not use data from other sources. This means that the scenario, where only one source is available, is processed in a similar way with the same detection performance.

For each case, simulations are performed with COFC of different frequencies (0.02, 0.2, and 2 Hz) and normalized amplitudes (0.2 and 0.8). For the single failure case the OFC amplitude is 0.2, and for the simultaneous OFC case the OFC is 0.8. These amplitudes correspond to the minimal OFC amplitude to detect for each case study in this paper. The objective arbitrarily chosen in this case study, for the two considered cases, is to carry out the OFC detection in less than three periods, with no false alarms. The corresponding detection delays are 150, 15, and 1.5 s. In the general case, depending on the considered aircraft and flight parameter, if the detection delay limits are not satisfied, additional structural loads, beyond the specified envelope, could be generated because of the OFC.

The tuning parameters of the Kalman filter are $\omega_0 = 0.4\pi = 0.2$ Hz and $\xi = 0.9$. The parameters of the harmonic filter are set to $K = 20$ and $\alpha = 10$. The parameter ω_0 is taken as the middle of the frequency band of the expected OFCs, *i.e.*, 0.02–2 Hz. The parameters ξ, K, and α are chosen using a Pareto-optimum approach, maximizing fault detection ratio and minimizing detection delay, missed detections, and false alarms. Table 5.2 gives an idea on the selection procedure. The values are given for a COFC of amplitude 0.8 and frequency of 0.2 Hz.

For the robust derivative estimator, β is set to 1. If the measurement noise is weak, one can take $\Delta T/z.epsi = 10$, but for this application the ratio is fixed to 0.00154 to obtain the best detection/false alarm ratio. This value is obtained using the approach from the previous paragraph. For the slope change detector and the gradient-based approach, 0.3 s time window is used for computations. The choice was made to match the fastest OFC.

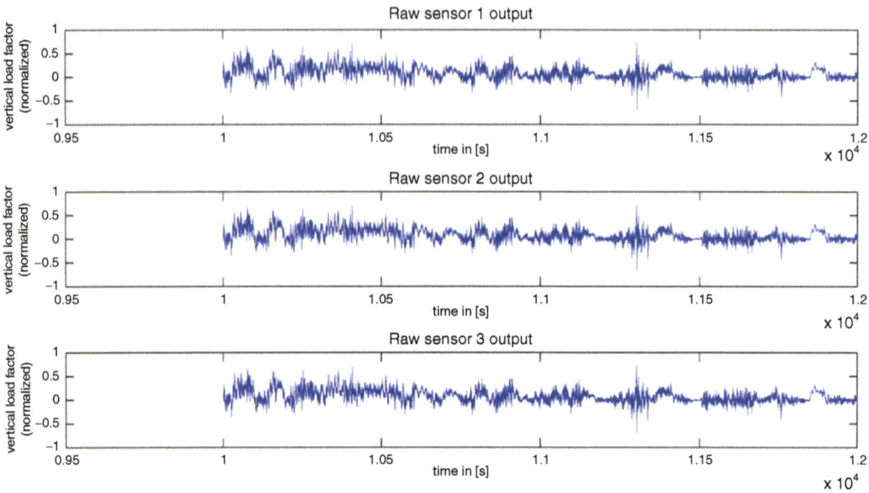

Fig. 5.11 Normalized redundant inertial data

5.4.2 *Single OFC Scenario*

Figure 5.11 shows the raw measurements provided by the three sources. Figure 5.12 shows the simulation results for the worst OFC case (OFC occurring in the first sensor, $a = 0.2, f = 0.02$ Hz). Threshold-based detection triggers if the count exceeds the threshold (100). Soft-computing-based FDI scheme (OFC detection in the table) triggers if a membership degree changes periodically from $0 \rightarrow 1$ and $1 \rightarrow 0$ four times. The detection delays are given in Table 5.3. The delays represent the difference between OFC occurrence in the system and OFC detection by the implemented FDI method. Threshold-based approach provides reasonably fast fault detection ($\approx$10 s) which is sufficient for 0.02 and 0.2 Hz OFCs but insufficient for 2 Hz. This is due to OFC related counter evolution: the count increases and decreases periodically, delaying the detection. Also, threshold-based approach does not discriminate between the different possible failures. Indeed, an OFC will not be distinguishable from a drift.

At the first glance, soft-computing-based approach exceeds the maximal delay by 50 s for the worst case, but the actual detection is performed by threshold crossing at 8.24 s, and the OFC is isolated at 200.4 s. Also, the approach shows remarkable speed for higher OFC frequencies and enables OFC discrimination before threshold-based detection. OFC can be detected and confirmed in less than three oscillations as required in this example.

Table 5.4 shows the delays obtained by the FDI module. The worst-case OFC is not detected, but the RDE performs well in the other cases; even if the detection

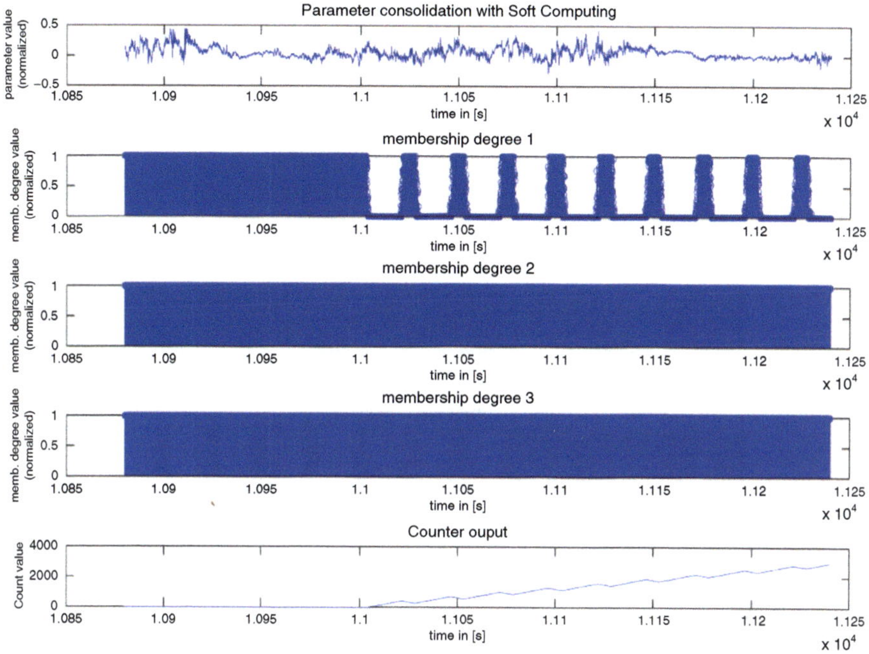

Fig. 5.12 Soft-monitoring simulation results for a single OFC

Table 5.3 Soft-monitoring detection delays

OFC	0.2, 0.02 Hz	0.2, 0.2 Hz	0.2, 2 Hz
Threshold-based det. (s)	8.24	8.6	8.6
OFC det. (s)	204.24	7.92	1.84

Table 5.4 OFC detection delays (in seconds)

	OFC		
Methods	0.2, 0.02 Hz	0.2, 0.2 Hz	0.2, 2 Hz
Gradient	X	X	0.48
Rob. derivative	X	37.72	1.16
Slope change det.	X	X	1.32

delay is exceeded for 0.2 Hz, the other approaches do not detect at all. In fact, the RDE-based approach can detect a 0.02 Hz frequency, 1.1 amplitude OFC with a delay of 150 s and an OFC of 0.03 Hz frequency and 0.55 amplitude in less than 140 s. In all the failure cases, the effect of the failure is compensated by the data consolidation module (based on soft voting).

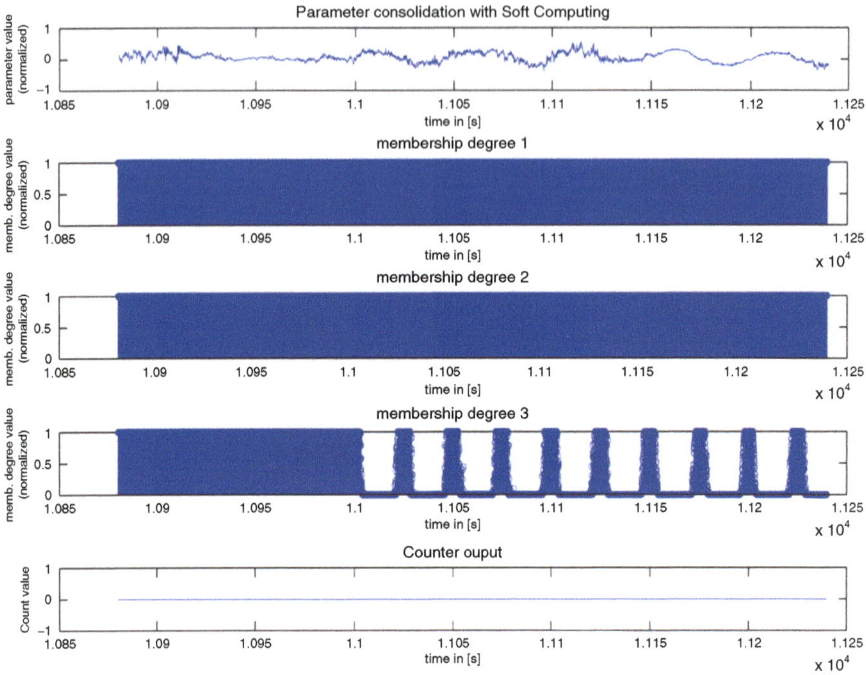

Fig. 5.13 Soft-monitoring simulation results for COFC

5.4.3 Combined OFC Scenario

When COFCs (simultaneous OFCs) appear on the first and the second sensors, the consolidation module succeeds detection (with detection delays similar to those given in Table 5.3), but fails to isolate and to switch off the faulty sources. As shown in Fig. 5.13, the two faulty sources possess a high membership degree and give a low membership degree to the valid third source. As a result the third source measurements are discarded, and the consolidated parameter is computed using faulty data.

On the other hand, fault isolation is successfully carried out by the FDI module. This is an important improvement in the context of aircraft structure optimization. Figures 5.14, 5.15, and 5.16 show the simulation results for the worst COFC case ($a = 0.8$, $f = 0.02$ Hz) for the three considered techniques: gradient, robust derivative estimator, and slope change detection. Again, RDE-based approach shows better robustness to noise when dealing with low-frequency OFCs, even if OFC isolation delay is important. As mentioned before, the minimal OFC amplitude detectable at a 0.02 Hz frequency is 1.1.

Table 5.5 shows the results for the remaining COFC cases. The values listed are for the first and the second sensors' OFC detection if the delays are different. Note that the implemented gradient-based and SCD approaches fail to detect the COFC

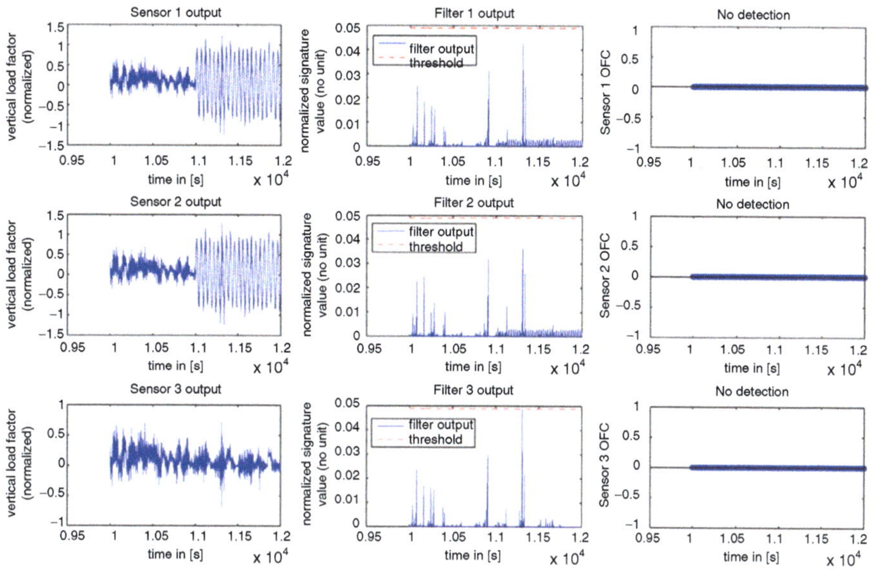

Fig. 5.14 COFC detection and isolation based on gradient

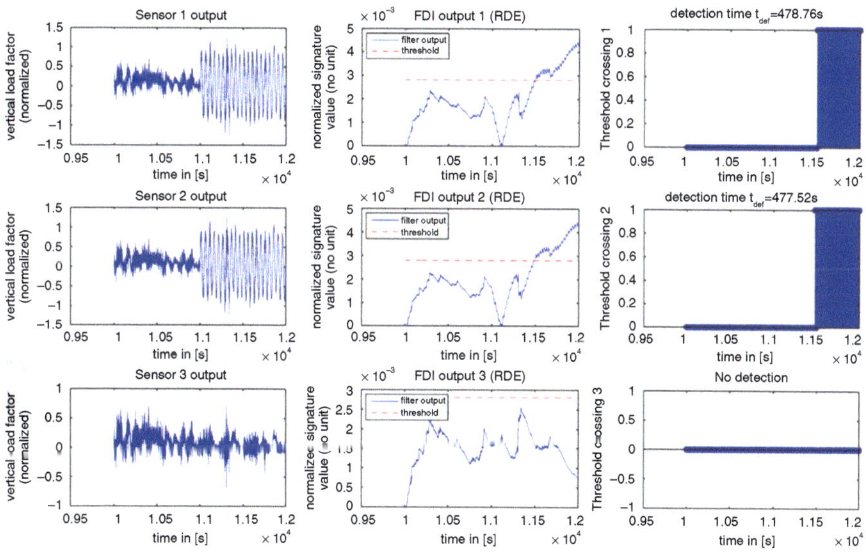

Fig. 5.15 COFC detection and isolation based on robust derivative estimator

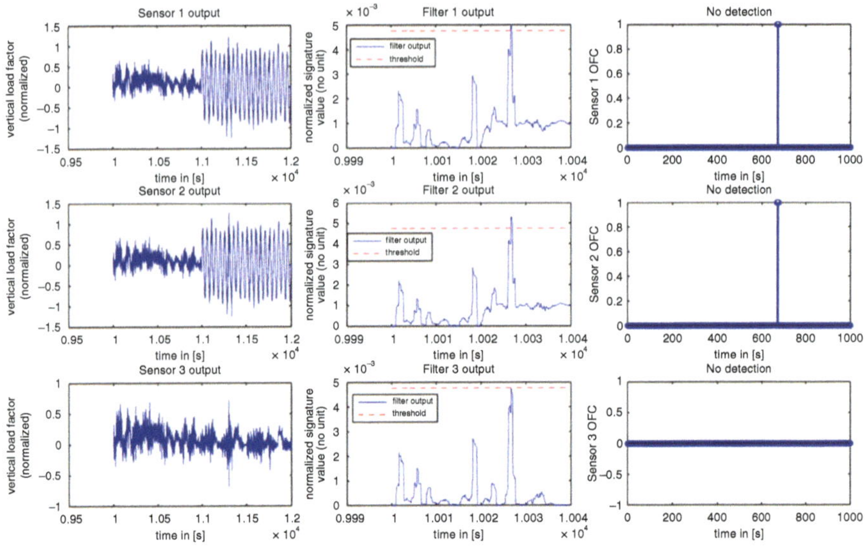

Fig. 5.16 COFC detection and isolation based on slope change detection

Table 5.5 COFC detection delays (in seconds)

	OFC		
Methods	0.8, 0.02 Hz	0.8, 0.2 Hz	0.8, 2 Hz
Gradient	X	0.64	2.2
Rob. derivative	478.76 and 477.52	2.84	2.2
Slope change det.	X	1.32	1.32

of the lowest frequency. However, the second and the third methods give satisfactory results: even if the detection delays seem to be important, these results are acceptable for this case. Note that the delay for the slope change detection method cannot be lower than 1.32 s because of the chosen time window.

5.4.4 Other Faults

The OFC indicator generator module is tuned to be sensitive to sinusoidal signals and to ignore the rest. Figure 5.17 confirms that behavior: a drift is occurring on sensor 1, and the OFC indicator shows no reaction. At the same time, the soft monitor reacts to the failure with a delay of 13.28 s (see Fig. 5.18, with definitions given in Sect. 5.4.2). This shows the benefit of the specialized modules: the soft monitor and the OFC indicators contributions are complementary.

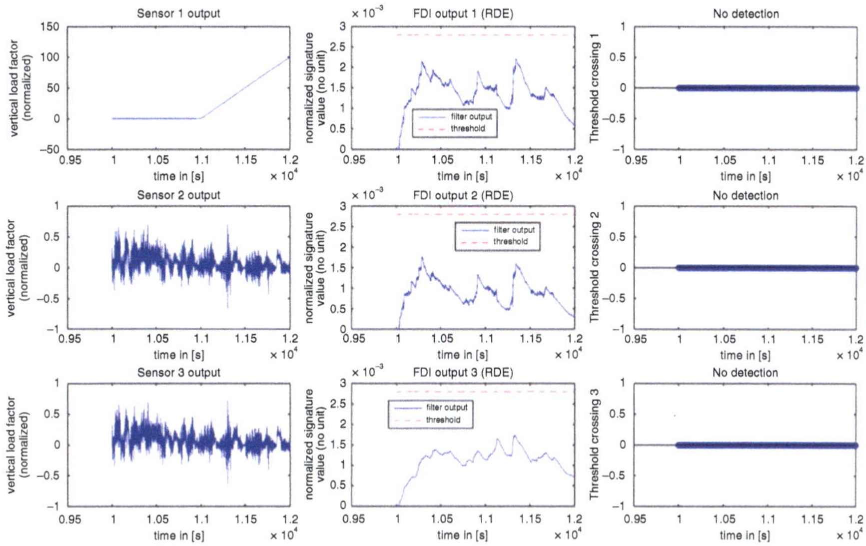

Fig. 5.17 OFC indicators behavior facing sensor drift failure

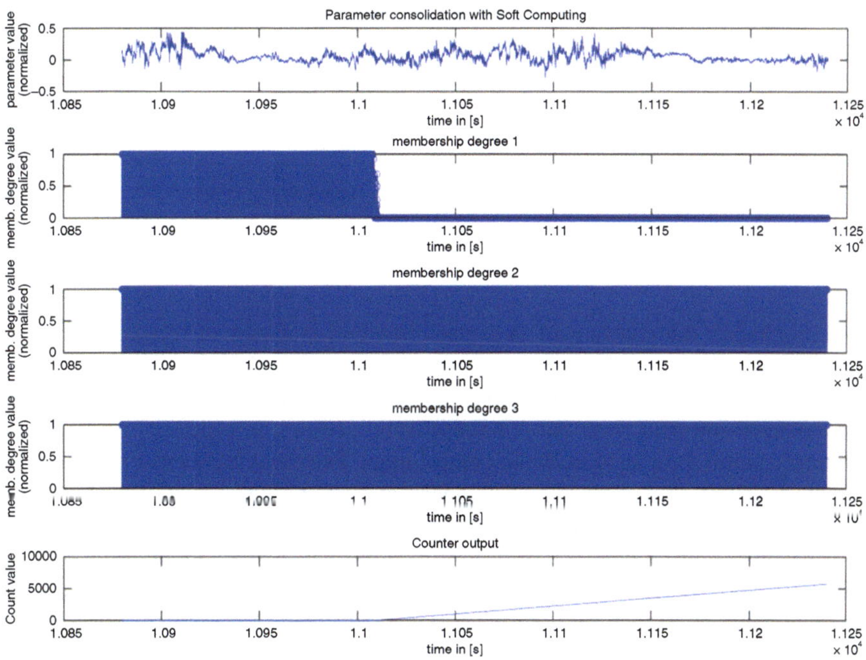

Fig. 5.18 Soft monitor "drift" failure detection

5.4.5 *Discussion*

Simulation results show the benefits of a combined soft-monitoring and dedicated OFC detection. This DF/FDI approach handles a broader range of failures and in particular OFCs of low frequencies with acceptable detection delays for the case study chosen for comparison purpose. When all flight parameter sensors are healthy, soft monitoring successfully detects and switches off a single corrupted source, using threshold-based and OFC-specific detection to cover drifts, freezing, and OFCs. When the soft monitoring is not suitable, harmonic filtering FDI successfully detects COFC failures using appropriate abrupt change detection methods. Another possible solution to this problem would be a bank of dedicated FDI modules, each module designed to handle a particular failure. From a system integration point of view, it would be better to integrate a dedicated module, within a more general module that handles the rest of the problems, like detection of simple failures and parameter consolidation. The RDE approach appears to be the most robust approach to detect OFCs, since it outperforms the other methods at low OFC frequencies and gives satisfactory results for the remaining OFCs. There is another aspect: Fig. 5.19 shows the three OFC indicators – gradient, RDE, and SCD – for the same OFC. It appears that the gradient and the SCD approaches are more sensitive to the noise than RDE. This makes the computation of an appropriate threshold difficult. A straightforward solution is to use additional filtering (Fig. 5.20) at the expense of additional delay and processing time. If frequencies superior to 0.2 are considered, SCD and gradient-based approaches are faster. To improve the performance that can be achieved for COFCs of low amplitudes and frequencies, further investigation will be necessary for optimal tuning of the harmonic filter.

5.5 Conclusion

In this chapter the problem of fault detection and isolation of redundant aircraft sensors, which are used for flight control laws computation, is investigated. The objective is to switch off the erroneous sensor and to compute a consolidated parameter using data from the remaining valid sensors, in order to eliminate any anomaly before propagation in the control loop. The focus has been on oscillatory failures in flight parameter, like, e.g., anemometric and inertial data. The proposed solution is based on a hierarchical monitoring scheme which is composed of a soft-computing module for data consolidation and overall monitoring, and an auxiliary OFC detection and isolation module designed to supplement the first module when more than two sensors become corrupted. A harmonic filter enables to transform the oscillation detection problem into an abrupt change detection problem. Three different evaluation approaches were tested to select the optimal performance/complexity trade-off, and the robust derivative estimator gave the best overall results. Also, a zero-crossing method is proposed to isolate OFCs using the soft monitor. The proposed approach was successfully tested on normalized data sets

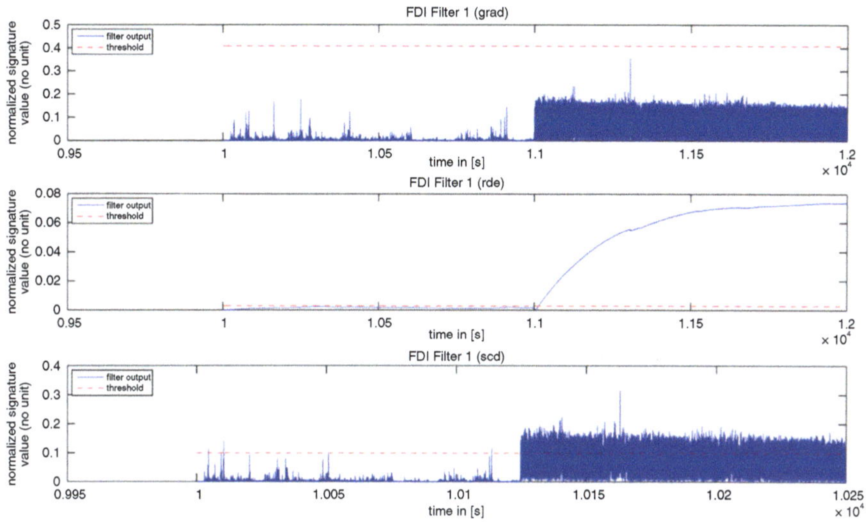

Fig. 5.19 OFC indicators without additional filtering

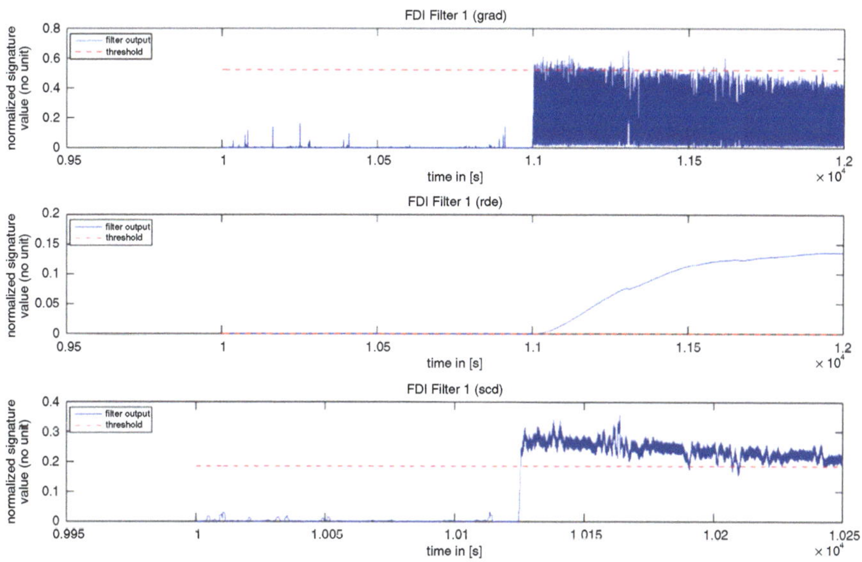

Fig. 5.20 OFC indicators with additional filtering

recorded from a real flight test, confirming the benefits of the combined FDI method. However, simulation results also show issues that can be addressed in future works. For example, the optimization of the harmonic filtering with FDI constraints will help dealing with low amplitude COFCs in noisy conditions. Also, the problem of dealing with multiple non-OFCs failures is very challenging.

References

1. Steffen T (2005) Control reconfiguration of dynamical systems: linear approaches and structural tests, Lectures notes in control and information sciences. Springer, Berlin
2. Allerton D, Jia H (2008) Distributed data fusion algorithms for inertial network systems. Radar Sonar Navig IET 2(1):51–62
3. Oosterom M, Babuska R, Verbruggen H (2002) Soft computing applications in aircraft sensor management and flight control law reconfiguration. IEEE Trans Syst Man Cybern Part C Appl Rev 32(2):125–139
4. Hegg J (2002) Enhanced space integrated GPS/INS (SIGI). IEEE Aerosp Electron Syst Mag 17(4):26–33
5. Kerr TH (1987) Decentralized filtering and redundancy management for multisensor navigation. IEEE Trans Aerosp Electron Syst AES-23(1):83–119
6. Lawrence P, Berarducci M (1996) Navigation sensor, filter, and failure mode simulation results using the distributed Kalman filter simulator (DKFSIM). In: Proceedings of IEEE PLANS, Atlanta, Georgia, April 22–26, pp 697–710
7. Rao B, Durrant-Whyte H (1991) Fully decentralised algorithm for multisensory Kalman filtering. IEEE Proc Control Theory Appl 138(5):413–420
8. Tupysev V (2000) Federated Kalman filter via formation of relation equations in augmented state space. J Guid Control Dyn 23(3):391–398
9. Boskovic JD, Mehra RK (2002) Failure detection, identification and reconfiguration in flight control. In: Fault diagnosis and fault tolerance for mechatronic systems: recent advances. Springer, Berlin
10. Alcorta-garcia E, Zolghadri A, Goupil P (2011) A nonlinear observer-based strategy for aircraft oscillatory failure detection: A380 case study. IEEE Trans Aerosp Electron Syst 47:2792–2806
11. Goupil P (2010) Oscillatory failure case detection in the A380 electrical flight control system by analytical redundancy. Control Eng Pract 18(9):1110–1119
12. Berdjag D, Cieslak J, Zolghadri A (2012) Fault diagnosis and monitoring of oscillatory failure case in aircraft inertial system. Control Eng Pract 20(12):1410–1425
13. Hagglund T (1995) A control loop performance monitor. Control Eng Pract 3(11):1543–1551
14. Thornhill NF, Hagglund T (1997) Detection and diagnosis of oscillation in control-loop. Control Eng Pract 5(10):1343–1354
15. Miao T, Seborg DE (1999) Automatic detection of excessively oscillatory feedback in control loops. In: IEEE international conference on control applications, Hawaii, pp 359–364
16. Thornhill NF, Huang B, Zhang H (2003) Detection of multiple oscillations in control loops. J Process Control 13(1):91–100
17. Harris TJ, Seppala CT, Desborough LD (1999) A review of performance monitoring and assessment techniques for univariate and multivariate control systems. J Process Control 9(1):1–17
18. Xia C, Howell J (2003) Loop status monitoring and fault localization. J Process Control 13(7):679–691
19. Shoukat Choudhury MAA, Shah SL, Thornhill NF (2004) Diagnosis of poor control-loop performance using higher-order statistics. Automatica 40:1719–1728
20. Kay S, Gabriel J (2002) Optimal invariant detection of a sinusoid with unknown parameters. IEEE Trans Signal Process 50(1):27–40
21. Yang ZY, Chan CW, Mok HT (2006) An approach to detect and isolate faults for nonlinear systems with periodic input. In: Proceedings of the SAFEPROCESS'06 conference, no. 1. Beijing, PR China, pp 301–306
22. Zivanovic M (2011) Detection of non-stationary sinusoids by using joint frequency reassignment and null-to-null bandwidth. Digit Signal Process 21(1):77–86
23. Odgaard PF, Wickerhauser MV (2007) Karhunen–Loeve based detection of multiple oscillations in multiple measurement signals from large-scale process plants. In: American control conference, New York City, USA, pp 5893–5898

24. Xia C, Howell J (2005) Isolating multiple sources of plant-wide oscillations via independent component. Control Eng Pract 13:1027–1035
25. Wilhelm L, Proetzsch M, Berns K (2009) Oscillation analysis in behavior-based robot architectures. In: Autonome mobile system. Springer, Berlin, p 121
26. Aranovskii SV, Bobtsov AA, Kremlev AS, Luk'yanova GV (2007) A robust algorithm for identification of the frequency of a sinusoidal signal. J Comput Syst Sci Int 46(3):371–376
27. Osder S (1999) Practical view of redundancy management application and theory. J Guid Control Dyn 22(1):12–21
28. Goupil P (2009) AIRBUS State of the art and practices on FDI and FTC. In: Proceedings of the 7th IFAC symposium on fault detection, supervision and safety of technical processes, Barcelona, Spain, July, pp 564–572
29. Odgaard PF, Trangbaek K (2006) Comparison of methods for oscillation detection – case study on a coal-fired power plant. In: Proceedings of the 5th IFAC symposium on power plants and power systems control, vol 5, Kananaskis, Canada, pp 297–302
30. Middleton R, Goodwin G (1990) Digital control and estimation. Prentice Hall, Inc., Englewood Cliffs
31. Zolghadri A (1996) An algorithm for real-time failure detection in Kalman filters. IEEE Trans Autom Control 41(10):1537–1539
32. Patton RJ (1994) Robust model-based fault diagnosis: the state of the art. In: Proceedings IFAC symposium SAFEPROCESS'94, vol. 1, Espoo, Finland, pp 1–24
33. Isermann R (2005) Model-based fault detection and analysis – status and application. Annu Rev Control 29:71–85
34. Frank P, Ding SX (1997) Survey of robust residual generation and evaluation methods in observer-based fault detection systems. J Process Control 7(6):403–424
35. Basseville M, Nikiforov IV (1993) Detection of abrupt changes – theory and application. Prentice-Hall, Inc., Englewood Cliffs
36. Marzat J, Piet-Lahanier H, Damongeot F, Walter E (2009) A new model-free method performing closed-loop fault diagnosis for an aeronautical system. In: 7th workshop on advanced control and diagnosis, ACD'2009. Zielona Gora, Poland. http://www.issi.uz.zgora.pl/ACD_2009/program/Papers/06_ACD_2009.pdfS
37. Dabroom A, Khalil H (1999) Discrete-time implementation of high-gain observers for numerical differentiation. Int J Control 72(17):1523–1537
38. Berdjag D, Zolghadri A, Cieslak J, Goupil P (2010) Fault detection and isolation for redundant aircraft sensors. In: Proceedings of the conference on control and fault-tolerant systems (SysTol'10), Nice, France, October, pp 137–142

Chapter 6
An Active Fault-Tolerant Flight Control Strategy

6.1 Introduction

6.1.1 Problem Statement

The problem studied in this chapter is that of design and analysis of an active fault-tolerant flight control system. The chapter presents a practical case study taken from the European GARTEUR[1] project (Flight Mechanics Action Group 16) on fault-tolerant control. Piloted flight simulator experiments are presented which show that fault tolerance can be achieved provided that there exists sufficient onboard control authority.

As briefly discussed in Chap. 2, the topic of FTC has recently received considerable attention [1–5]. For aircraft applications, the objective is to help the crew recover control capabilities quickly during a fault situation. FTC strategies can be classified into passive and active approaches. In the passive approach, the control algorithm is designed so that the system is able to achieve specified objectives, in fault-free as well as in fault situations. Obviously, guaranteed robustness to some a priori known faults is achieved at the expense of performance deterioration in the fault-free mode. In an active FTC system, faults are detected and identified by an FDD system, and the control laws are reconfigured accordingly online. The performance for the nominal and healthy operating mode is not degraded. Here, faults could be of a priori known type or could be unforeseeable ones. The latter case cannot be dealt with using a passive strategy. The reconfiguration mechanism is activated as soon as the FDD system detects and confirms the presence of a fault. Fault detection/confirmation should be fast and robust for

[1]Group for Aeronautical Research and Technology in EURope. See http://www.nlr.nl/documents/GARTEUR_AG16_Workshop/

A. Zolghadri et al., *Fault Diagnosis and Fault-Tolerant Control and Guidance for Aerospace Vehicles: From Theory to Application*, Advances in Industrial Control,
DOI 10.1007/978-1-4471-5313-9_6,

successful FTC. Obviously, the feasibility of both FTC approaches is dependent on the recoverability/compensability of each fault, a problem that, surprisingly, has received relatively little attention in the literature [6, 7]. In Chap. 2, it has been underlined that the case of non-compensable actuator faults for an RLV can be studied as a trimmability-deficiency analysis, i.e., the flight envelope regions where the vehicle cannot be rotationally balanced in the presence of faults. A fault is then considered as non-compensable if the flight trajectory of the vehicle crosses the non-trimmable region. The same fault compensability analysis will be applied in this chapter.

A great number of solutions for active FTC are available in the open literature. Design methodologies include those based on the linear quadratic control scheme [8, 9], modular approach [10], Model-Based Predictive Control (MPC) method [11, 12], and H_∞ control theory [13, 14], to name a few. Other works are based on Linear Parameter Varying (LPV) techniques [15, 16], where the idea is to use the output of the FDI scheme, jointly with some subspace of the system states, as scheduling parameters of the LPV fault-tolerant controller.

The above works offer attractive conceptual features. At the same time, a number of them present several shortcomings for effective in-flight implementation: significant increase in computational burden or drastic modification of the control system structure is already in place. Note also that some papers have developed control allocation mechanisms for redistributing the total control effort among the remaining healthy actuators (see, e.g., [17–19]).

In this chapter, the proposed FTC system operates in such a way that once a fault is detected and confirmed by the FDD system, a compensation loop is activated for safe recovery. A key feature is that the added FTC loop keeps unchanged the in-service control laws, facilitating the certification of the whole approach and limiting the underlying verification and validation activities. The solution could be a good and technologically viable candidate for effective fault-tolerant flight control. The material of this chapter is mostly underpinned by the published papers [7, 14]. For clarity of presentation, some proofs are omitted here. The interested reader can refer to those references for more details.

6.2 Fault-Tolerant Control Architecture

To illustrate the basic idea, consider a feedback loop as shown in Fig. 6.1, where G is the plant to be controlled and K_o is the nominal controller. Generally, the nominal controller is validated for fault-free situation.

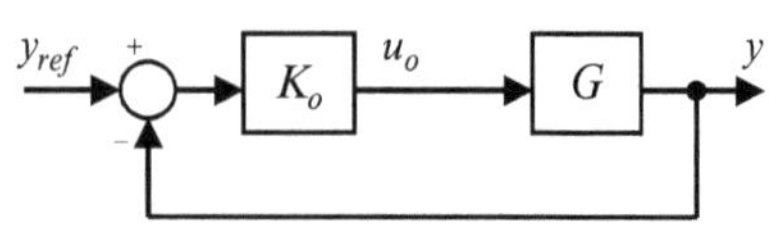

Fig. 6.1 Standard feedback configuration

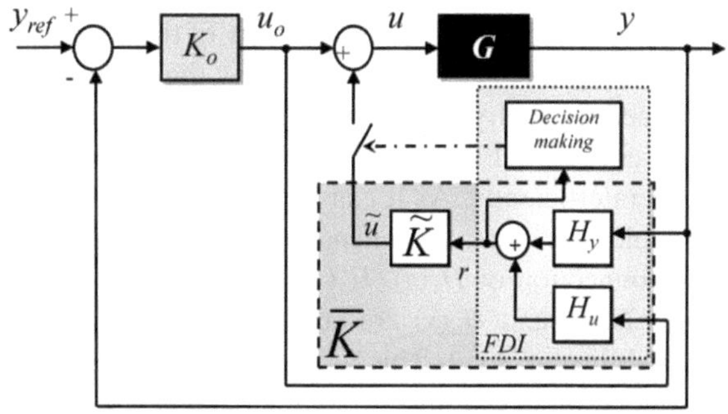

Fig. 6.2 General FTC setup with a model-based FDD scheme

After the occurrence of faults, the system dynamics can be affected, and the feedback control achieved by K_o may result in unsatisfactory performance and even instability. The main idea of the proposed solution is to reconfigure the control loop in such a way that the nominal controller K_o is still used without any retuning. To achieve fault tolerance, a fault-tolerant control loop is added to compensate the effects that the faults could have on the system, i.e., the input/output signals u_o and y seen by K_o have a "similar" dynamic behavior as for the fault-free situation. The "similarity" can be expressed in terms of a certain distance between the two behaviors. Of course, such an index is highly connected to the nature and severity of considered faults.

6.2.1 *FTC with a Model-Based FDD Scheme*

Following the basic ideas presented in [13], we propose to design of the FTC loop according to the block diagram shown in Fig. 6.2. The reconfigurable recovery scheme is divided into three parts:

- FDD unit represented by two linear filters $H_y(s)$, $H_u(s)$ which generates continuously a fault indicating signal r:

$$r(s) = H_u(s)u_0(s) + H_y(s)y(s).$$

  We assume that simple threshold logic is used to detect and confirm the presence of faults through r.
- FTC part represented by $\tilde{K}(s)$ which generates an additional control signal $\tilde{u}$ to be added to the nominal control signal u_o in a faulty situation.
- FTC activation mechanism to activate the FTC strategy.

Once again, in a fault-free situation, the FTC loop is not activated leaving the plant only controlled by the nominal controller. When the FTC strategy is activated, the control law is reconfigured by adding the signal $\tilde{u}$ to the nominal control signal u_o. Since the activation of this loop is done by using a switching logic, the overall scheme keeps nominal flight performance in fault-free situations.

This proposed active FTC architecture implies some important issues. The first question concerns the activation delay of the FTC strategy. During this time interval, the faulty system is still controlled by the nominal control law which has not been designed for faulty situations. This problem is also highly related to the time delay detection of the FDI part. Some solutions are discussed in [14, 20] to address this problem efficiently. As it can be seen in Fig. 6.2, the FTC scheme is in open loop for fault-free situations. Then, an important requirement for FTC scheme is that the interconnection of $H_y(s)$, $H_u(s)$, and $\tilde{K}(s)$ depicted from Fig. 6.2 must be stable. Since $H_y(s)$ and $H_u(s)$ are stable detection filters, this problem is equivalent to the stability of $\tilde{K}(s)$. This will be discussed and clarified in Sect. 6.2.4. The FTC design problem can thus be summarized as the design of a dynamical fault-tolerant controller $\tilde{K}(s)$ that allows for "input/output insensitivity" despite the presence of the fault, i.e.:

Problem 6.1 Suppose that the fault is recoverable/compensable [6, 7]. The goal is to design a stable controller $\tilde{K}(s)$ to produce the new control signal:

$$u(s) = u_o(s) + \tilde{K}(s)r(s) \tag{6.1}$$

such that the stability of the feedback system and the required control specifications are guaranteed for considered faults. Using an H_∞ formulation [21, 22], this means that $\tilde{K}(s)$ should satisfy the constraint:

$$\left\| F_l\left(P_1(s), \tilde{K}(s)\right)\right\|_\infty < \gamma_1 \tag{6.2}$$

where $P_1(s)$ is deduced from $K_o(s)$, $G(s)$, $H_y(s)$, and $H_u(s)$ using some standard algebraic manipulations. The scalar γ_1 denotes some FTC performance level to be achieved. In this formulation, $F_l\left(P_1(s), \tilde{K}(s)\right)$ corresponds to the lower LFT (linear fractional transformation) of $P_1(s)$ by $\tilde{K}(s)$. ■

6.2.2 FTC with Dedicated Onboard FDD

In this section, we consider that FDD task relies on existing monitoring systems and the fault indicating output is available onboard. FTC problem boils down to the design of $\tilde{K}(s)$ (Fig. 6.3). The information of the FDD unit is used to activate the fault accommodation. In this case, the synthesis problem can be formulated as follows:

Problem 6.2 Suppose that the fault is recoverable/compensable [6, 7]. The goal is to design a stable controller $\overline{K}(s)$ to produce the new control signal:

$$\tilde{u}(s) = u_o(s) + \overline{K}(s)\varepsilon(s) \tag{6.3}$$

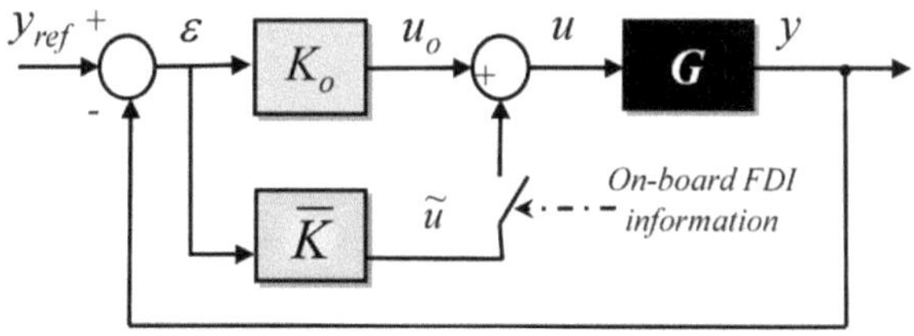

Fig. 6.3 General FTC setup with an onboard FDI scheme

such that the stability of the feedback system and the required control specifications are guaranteed for considered faults. Using an H_∞ formulation [21, 22], this means that $\overline{K}(s)$ should satisfy the constraint:

$$\left\| F_l\left(P_2(s), \overline{K}(s)\right)\right\|_\infty < \gamma_2 \tag{6.4}$$

where $P_2(s)$ is deduced from $K_o(s)$ and $G(s)$ after some linear fractional algebra manipulations. γ_2 represents some performance level to achieve. ■

Remark 6.1 In Figs. 6.2 and 6.3, it is natural to ask about the stability of the FTC loop in the presence of the switch. Here, we assume that once a fault is detected, the switch is definitively activated and the compensation signal $\tilde{u}$ remains active. That is solution deals with strongly detectable faults (the effect of a fault persist on the corresponding fault indicating signal) and the remaining problem concerns the transient behavior of $\tilde{u}$. To avoid "bumps," a solution to manage this problem is given in Sect. 6.3. The case of intermittent faults can be dealt with, for example, by a supervisory FTC setup [23–29].

6.2.3 Analysis of FTC Architecture

In this section, the FTC setup with a model-based FDD scheme is analyzed to highlight some interesting features with respect to the interaction between the FDI and FTC units. The goal is to derive some assumptions about the FDI schemes for an integrated FDI/FTC design approach.

Consider the block diagram shown in Fig. 6.2. Let (A, B, C, D), $\left(\tilde{A}, \tilde{B}, \tilde{C}, \tilde{D}\right)$, (A_u, B_u, C_u, D_u), and (A_y, B_y, C_y, D_y) be the state-space representation of $G(s)$, $\tilde{K}(s)$, $H_u(s)$, and $H_y(s)$, respectively. The state-space model $G_{FTC}(s)$, which represents the dynamic channel between the nominal control signal u_o and the measurements y, is derived from $G(s)$, $\tilde{K}(s)$, $H_u(s)$, and $H_y(s)$ according to

$$G_{\text{FTC}}: \begin{cases} \begin{pmatrix} \dot{x}_c \\ \dot{x}_u \end{pmatrix} = \begin{pmatrix} A_{11} & A_{12} \\ 0 & A_u \end{pmatrix}\begin{pmatrix} x_c \\ x_u \end{pmatrix} + \begin{pmatrix} B_1 \\ B_u \end{pmatrix} u_o \\ y = \begin{pmatrix} C_1 & C_2 \end{pmatrix}\begin{pmatrix} x_c \\ x_u \end{pmatrix} + D_{22} u_o \end{cases}. \tag{6.5}$$

The matrix A_{11}, A_{12}, B_1, C_1, C_2, and D_{22} can be deduced from the above state-space representations as follows:

$$A_{11} = \begin{pmatrix} A + BM\tilde{D}D_yC & BM\tilde{C} & BM\tilde{D}C_y \\ \tilde{B}D_y\left(C + DM\tilde{D}D_yC\right) & \tilde{A} + \tilde{B}D_yDM\tilde{C} & \tilde{B}\left(I + D_yDM\tilde{D}\right)C_y \\ B_y\left(I + DM\tilde{D}D_y\right)C & B_yDM\tilde{C} & A_y + B_yDM\tilde{D}C_y \end{pmatrix},$$

$$A_{12} = \begin{pmatrix} BM\tilde{D}C_u \\ \tilde{B}\left(I + D_yDM\tilde{D}\right)C_u \\ B_yDM\tilde{D}C_u \end{pmatrix}, B_1 = \begin{pmatrix} BM(I + \tilde{D}D_u) \\ \tilde{B}\left(D_u + D_yDM(I + \tilde{D}D_u)\right) \\ B_yDM(I + \tilde{D}D_u) \end{pmatrix},$$

$$C_1 = \left(C + DM\tilde{D}D_yC \;\; DM\tilde{C} \;\; DM\tilde{D}C_y\right), C_2 = \left(DM\tilde{D}C_u\right),$$

$$D_{22} = DM\left(I + \tilde{D}D_u\right), M = \left(I - \tilde{D}D_yD\right)^{-1}.$$

The augmented state vector x_c is given by $x_c = \left(x^{\mathrm{T}} \; \tilde{x}^{\mathrm{T}} \; x_y{}^{\mathrm{T}}\right)^{\mathrm{T}}$, where x, $\tilde{x}$, x_y, and x_u are the state vectors associated with $G(s)$, $\tilde{K}(s)$, $H_y(s)$, and $H_u(s)$, respectively.

From (6.5), it can be seen that the poles of $G_{\mathrm{FTC}}(s)$ are given by the eigenvalues of A_{11} and A_u. Note that the expression for A_{11} does not contain the A_u, B_u, C_u, and D_u matrices. It follows that $H_u(s)$ (stable filter) does not impact the stability of $G_{\mathrm{FTC}}(s)$.

Now, consider the overall FTC architecture described in Fig. 6.2, and let the state-space representations of $K_o(s)$ and $G_{\mathrm{FTC}}(s)$ be given by (A_o, B_o, C_o, D_o) and (A_G, B_G, C_G, D_G), respectively. By definition

$$A_G = \begin{pmatrix} A_{11} & A_{12} \\ 0 & A_u \end{pmatrix}, B_G = \begin{pmatrix} B_1 \\ B_u \end{pmatrix}, C_G = \left(C_1 \; C_2\right), D_G = D_{22}.$$

Let x_o be the state vector of $K_o(s)$ and denote by x_G the augmented vector so that $x_G = \left(x^{\mathrm{T}} \; \tilde{x}^{\mathrm{T}} \; x_y{}^{\mathrm{T}} \; x_u{}^{\mathrm{T}}\right)^{\mathrm{T}}$. Direct calculations lead to the following closed-loop state-space model

$$\begin{cases} \begin{pmatrix} \dot{x}_G \\ \dot{x}_o \end{pmatrix} = A_{\mathrm{T}} \begin{pmatrix} x_G \\ x_o \end{pmatrix} + B_{\mathrm{T}} y_{\mathrm{ref}} \\ y = C_{\mathrm{T}} \begin{pmatrix} x_G \\ x_o \end{pmatrix} + D_{\mathrm{T}} y_{\mathrm{ref}} \end{cases} \tag{6.6}$$

where A_T, B_T, C_T, and D_T are given by

$$A_T = \begin{pmatrix} A_G - B_G D_o N C_G & B_G C_o - B_G D_o N D_G C_o \\ -B_o N C_G & A_o - B_o N D_G C_o \end{pmatrix},$$

$$B_T = \begin{pmatrix} B_G D_o (I - N D_G D_o) \\ B_o (I - N D_G D_o) \end{pmatrix},$$

$$C_T = \begin{pmatrix} N C_G & N D_G C_o \end{pmatrix}, D_T = (N D_G D_o), N = (I + D_G D_o)^{-1}. \tag{6.7}$$

Expression (6.6) shows that the stability of the overall loop depends on the stability of the FDI filter. This is an expected and rather evident result. Then, expression (6.6) suggests that the FDI and FTC performances are highly coupled.

A nice feature of the FTC architecture presented in Fig. 6.2 is that the $\overline{K}(s)$ block can be seen as the set of all admissible FDI/FTC units which achieve some level of performance. This suggests the following design procedure. Firstly assume that $\overline{K}(s)$ is designed according to some FTC specifications. Now, the problem is to deduce from $\overline{K}(s)$ the FDI part ($H_y(s)$ and $H_u(s)$) and the FTC part, $\tilde{K}(s)$. The proposed procedure consists of designing $H_y(s)$ and $H_u(s)$ and then to integrate the FDI performance specifications into the FTC design procedure. Thus, the computed FDI/FTC couple is a solution to the problem of integrated FTC/FDI unit design, if and only if this couple belongs to the set $\overline{K}(s)$, that is, if

$$\left\| F_l\left(P_2(s), F_l\left(F(s), \tilde{K}(s)\right)\right) \right\|_\infty < \gamma_2, \quad F(s) = \begin{pmatrix} H_y(s) & H_u(s) \end{pmatrix}. \tag{6.8}$$

The interested reader can refer to [30, 31] to find procedures where FDI filters are extracted from $\overline{K}(s)$.

6.2.4 Formulation of FTC Design

The main objective of the added FTC loop is to make input/output signals u_o and y seen by K_o exhibit as similar as possible dynamic behavior than for the fault-free situation. In this work, this "similarity" is expressed in terms of a certain distance between the two behaviors. To preserve stability and nominal performances, $G_{\text{FTC}}(s)$ must be close to $G(s)$ according to some metric. The goal is thus to design $H_y(s)$, $H_u(s)$, and $\tilde{K}(s)$ (or equivalently $\overline{K}(s)$) so that

$$\min_{(H_y(s), H_u(s), \tilde{K}(s))/\overline{K}(s)} \mathcal{M}(G_{FTC}(s), G(s)) \tag{6.9}$$

where $\mathcal{M}(.)$ denotes a specified metric.

Here, the above problem is addressed within the H_∞ mixed-sensitivity setting [21, 32] according to the following proposition.

Proposition 6.1 *Consider the block diagrams depicted in Figs.* 6.2 *and* 6.3. *Let S, R, and T denote the (nominal) sensitivity function, the sensitivity function of the controlled input and the complementary sensitivity function respectively, i.e.,*

$$\begin{aligned} S(s) &=(I + G(s)K_o(s))^{-1}, \\ R(s) &=K_o(s)(I + G(s)K_o(s))^{-1}, \\ T(s) &=G(s)K_o(s)(I + G(s)K_o(s))^{-1}. \end{aligned} \tag{6.10}$$

Denote the faulty sensitivity function $S_{\mathrm{FTC}}(s)$, *the faulty sensitivity function of the controlled input* $R_{\mathrm{FTC}}(s)$ *and let the faulty complementary sensitivity function* $T_{\mathrm{FTC}}(s)$, *be defined according to* (6.10) *by substituting* $G(s)$ *by* $G_{\mathrm{FTC}}(s)$. *Denote by* $W_1(s)$, $W_2(s)$, *and* $W_3(s)$ *the weighting functions used to shape* $S_{\mathrm{FTC}}(s)$, $R_{\mathrm{FTC}}(s)$, *and* $T_{\mathrm{FTC}}(s)$, *respectively. These functions should be defined according to*

$$\min_{W_1(s),W_2(s),W_3(s)} \left(\| G(s) \|_\infty - \| G_{\mathrm{FTC}}(s) \|_\infty \right). \tag{6.11}$$

Then, a necessary and sufficient condition for the FTC loop composed by $H_y(s)$, $H_u(s)$, *and* $\tilde{K}(s)$ *(or equivalently* $\overline{K}(s)$*) to preserve stability and performance is*

$$\bar{\sigma}(S_{\mathrm{FTC}}(j\omega)) \leq \bar{\sigma}({W_1}^{-1}(j\omega)), \quad \forall\omega, \tag{6.12}$$

$$\bar{\sigma}(R_{\mathrm{FTC}}(j\omega)) \leq \bar{\sigma}({W_2}^{-1}(j\omega)), \quad \forall\omega, \tag{6.13}$$

$$\bar{\sigma}(T_{\mathrm{FTC}}(j\omega)) \leq \bar{\sigma}({W_3}^{-1}(j\omega)), \quad \forall\omega, \tag{6.14}$$

where $\bar{\sigma}(\bullet)$ *denotes the maximal singular value of '•'. The gap between* $\bar{\sigma}({W_1}^{-1}(j\omega))$, $\bar{\sigma}({W_2}^{-1}(j\omega))$, *and* $\bar{\sigma}({W_3}^{-1}(j\omega))$ *and* $\bar{\sigma}(S_{\mathrm{FTC}}(j\omega))$, $\bar{\sigma}(R_{\mathrm{FTC}}(j\omega))$, *and* $\bar{\sigma}(T_{\mathrm{FTC}}(j\omega))$ $\forall\omega$ *indicate the loss of performance with respect to the nominal case.* □

Proof. Immediate application of the well-known mixed-sensitivity theory [32] to the above FTC problem. □

Proposition 6.1 provides a necessary and sufficient condition. However, as it has been outlined in Sect. 6.2.1, FTC open loop must be stable since it operates in open loop in fault-free situations. A necessary and sufficient condition for the existence of a stable stabilizing (*strong stabilization*) FTC loop for a given plant is the so-called parity interlacing property (PIP) [33] applied to $\tilde{K}(s)$ (or equivalently $\overline{K}(s)$). Since the H_∞ controller is in general not unique, it is reasonable to expect that even if the H_∞ central controller is unstable, there might still be a stable controller that could satisfy the H_∞ norm bound when the PIP condition is satisfied. This problem is studied in [33]. In the H_∞ "sensitivity mixed" context, the solution of this design problem can be formulated as follows. Note that the following proposition is given for $\tilde{K}(s)$ (Fig. 6.2), but similar developments can be used to $\overline{K}(s)$ (Fig. 6.3).

Proposition 6.2 *Assume that a solution to the H_∞ "mixed-sensitivity" problem exists for a $\gamma < 1$, i.e., there exists $\tilde{K}(s) = F_l(\hat{K}(s), Q(s))$ with $Q \in \Re H_\infty$ and $\|Q\|_\infty < \gamma$ such that* (6.12), (6.13), *and* (6.14) *hold. $F_l(\hat{K}(s), Q(s))$ is the set of all controllers satisfying* (6.12), (6.13), *and* (6.14). *Then, there exists a solution to the H_∞ strong stabilization problem if and only if there exists $Q = (A_q, B_q, C_q, D_q)$ of some suitable order, with $\|Q\|_\infty < \gamma$, such that*

$$\tilde{A} = \begin{pmatrix} \hat{A} + \hat{B}_2\hat{S}^{-1}D_q\hat{C}_2 & \hat{B}_2\hat{S}^{-1}C_q \\ B_q\hat{R}^{-1}\hat{C}_2 & A_q + B_q\hat{R}^{-1}\hat{D}_{22}C_q \end{pmatrix} \tag{6.15}$$

is stable, where $\hat{S} = I - D_q\hat{D}_{22}$ and $\hat{R} = I - \hat{D}_{22}D_q$. The matrix $\tilde{A}$ is the evolution matrix of $\tilde{K}$ and $\hat{A}, \hat{B}_1, \hat{B}_2, \hat{C}_1, \hat{C}_2, \hat{D}_{11}, \hat{D}_{12}$, and $\hat{D}_{21}$ and $\hat{D}_{22}$ denote the state-space matrices of $\hat{K}$ such that

$$\hat{K}(s) = \left[\begin{array}{c|cc} \hat{A} & \hat{B}_1 & \hat{B}_2 \\ \hline \hat{C}_1 & \hat{D}_{11} & \hat{D}_{12} \\ \hat{C}_2 & \hat{D}_{21} & \hat{D}_{22} \end{array}\right] \tag{6.16}$$

□

Proof. Immediate application of the Youla parameterization [21]. □

This proposition shows that the PIP condition is equivalent to finding a suitable Youla parameter such that $\tilde{A}$ is stable and $\|Q\|_\infty < \gamma$. In particular, the central regulator $\tilde{K}(s) = F_l(\hat{K}(s), 0) = \hat{K}(s)$ is a suitable solution if $\hat{A}$ is stable.

6.3 Bumpless Scheme

The remaining problem concerns the transient behavior of the signal $\tilde{u}$. In order to avoid undesirable transient phenomena, a practically relevant solution is now given.

6.3.1 Solution with a Model-Based FDD Scheme

Figure 6.4 presents the proposed solution to manage undesired bumps. The aim is to drive $\tilde{K}(s)$ before the switch by a gain F_s, such that $\tilde{u} \mapsto 0$and $\eta \mapsto r$ according to

$$\begin{cases} \tilde{u}(s) = \tilde{K}(s)\eta(s) \\ \eta(s) = F_s \begin{bmatrix} \tilde{x}(s) \\ r(s) \end{bmatrix} \end{cases}. \tag{6.17}$$

η denotes the control signal of $\tilde{K}(s)$ before the switch and F_s the static gain.

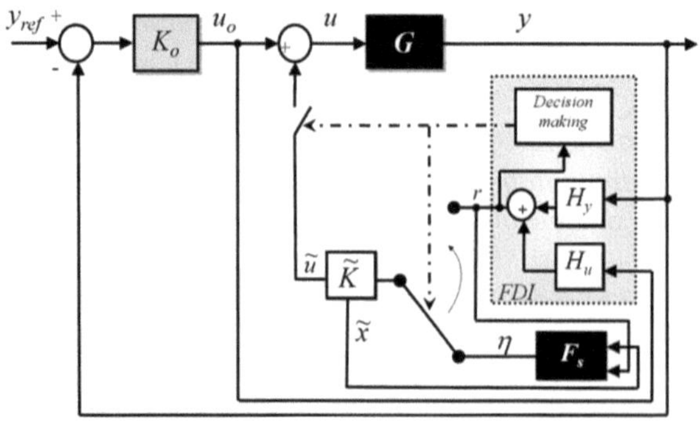

Fig. 6.4 FTC with model-based FDD and including bumpless mechanism

Different approaches can be used to design F_s. Here, we propose to use the idea initially proposed by Turner and Walker in [34]. To compute F_s, the following quadratic criterion is minimized:

$$J\left(\tilde{u},\eta\right)=\frac{1}{2}\int_0^{\infty}\left(\tilde{u}^T W_u\tilde{u}+\left(\eta-r\right)^T W_e\left(\eta-r\right)\right)\mathrm{d}t. \tag{6.18}$$

W_u and W_e are constant positive-definite weighting matrices of appropriate dimensions. W_u and W_e permit to define the desired objectives. For example, if it is desirable to minimize the magnitude of $\tilde{u}$, we should choose a high value for W_u. So, at switching time t_s (time where fault is detected), we have $\tilde{u}\left(t_s\right)\mapsto 0$, then $u\left(t_s\right)\mapsto u_o\left(t_s\right)$. Hence, there is no bump effect. Similarly, if we want to reduce the energy of $(\eta-r)$, the value of W_e must be set to a high value. Then, at t_s, we have $\eta\left(t_s\right)\mapsto r\left(t_s\right)$. So, there is no discontinuity between η and r at switching time. This means that from a practical point of view, a trade-off between minimizing the magnitude of $\tilde{u}$ and $(\eta-r)$ should be found.

Once W_u and W_e have been chosen, the solution is given by (the interested reader can refer to [34] for more details)

$$F_s=\bar{\Delta}\begin{bmatrix}\left(\tilde{B}^{\mathrm{T}}\Pi+\tilde{D}^{\mathrm{T}}W_u\tilde{C}\right)^{\mathrm{T}}\\ \left(-W_e+\tilde{B}^{\mathrm{T}}M\left(\tilde{C}^{\mathrm{T}}W_u\tilde{D}\bar{\Delta}W_e+\Pi\tilde{B}\bar{\Delta}W_e\right)\right)^{\mathrm{T}}\end{bmatrix}^{\mathrm{T}} \tag{6.19}$$

where M and $\bar{\Delta}$ are defined according to

$$M=\left(\mathsf{A}^{\mathrm{T}}+\Pi\mathsf{B}\right)^{-1} \tag{6.20}$$

$$\bar{\Delta}=-\left(\tilde{D}^{\mathrm{T}}W_u\tilde{D}+W_e\right)^{-1}. \tag{6.21}$$

The matrix Π is the definite-positive stationary solution of the following algebraic Riccati equation:

$$\Pi\mathsf{A}+\mathsf{A}^{\mathrm{T}}\Pi+\Pi\mathsf{B}\Pi+\mathsf{C}=0 \tag{6.22}$$

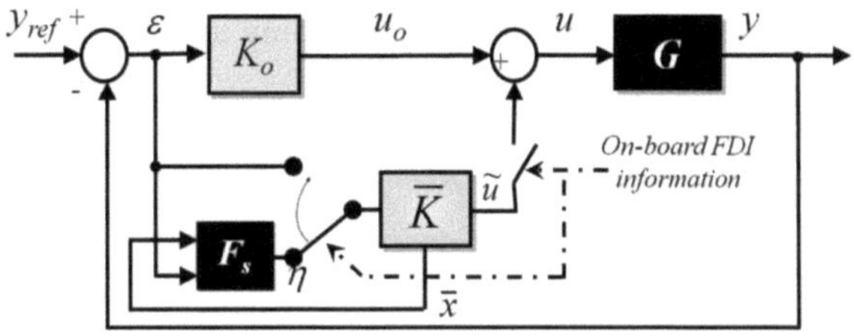

Fig. 6.5 FTC with available onboard FDD and including bumpless mechanism

The matrix A, B, and C are given by

$$\mathsf{A} = \tilde{A} + \tilde{B}\bar{\Delta}\tilde{D}^{\mathrm{T}} W_u \tilde{C} \tag{6.23}$$

$$\mathsf{B} = \tilde{B}\bar{\Delta}\tilde{B}^{\mathrm{T}} \tag{6.24}$$

$$\mathsf{C} = \tilde{C}^{T} W_u \left(I + \tilde{D}\bar{\Delta}\tilde{D}^{T} W_u\right) \tilde{C} \tag{6.25}$$

where $\left(\tilde{A}, \tilde{B}, \tilde{C}, \tilde{D}\right)$ denotes the state-space matrices of $\tilde{K}(s)$.

6.3.2 *Solution with Dedicated Onboard FDD*

Figure 6.5 presents the proposed solution to manage the transient behaviors for the FTC setup with an available FDD unit. In this case, the aim is to drive $\overline{K}(s)$ before the switch by a gain F_s, such that $\tilde{u} \mapsto 0$and $\eta \mapsto \varepsilon$ according to

$$\begin{cases} \tilde{u}(s) = \overline{K}(s)\eta(s) \\ \eta(s) = F_s \begin{bmatrix} \bar{x}(s) \\ \varepsilon(s) \end{bmatrix} \end{cases} \tag{6.26}$$

where η denotes now the control signal of $\overline{K}(s)$ before the switch, $\bar{x}$ the state vector of $\overline{K}(s)$, and F_s the gain to design.

Here, the quadratic criterion defined in (6.18) can be rewritten as follows:

$$J\left(\tilde{u}, \eta\right) = \frac{1}{2}\int_0^{\infty} \left(\tilde{u}^T W_u \tilde{u} (\eta - \varepsilon)^{\mathrm{T}} W_e \left(\eta - \varepsilon\right)\right) \mathrm{d}t \tag{6.27}$$

W_u and W_e are constant positive-definite weighting matrices of appropriate dimensions defined to select some desired objectives in a same way as in the Sect. 6.3.1. Once W_u and W_e have been chosen, the solution to FTC architecture depicted in Fig. 6.5 is given by

$$F_s = \bar{N} \begin{bmatrix} \left(\bar{B}^{\mathrm{T}}\Pi + \bar{D}^{\mathrm{T}} W_u \bar{C}\right)^{\mathrm{T}} \\ \left(-W_e + \bar{B}^{\mathrm{T}} \bar{M}\left(\bar{C}^{\mathrm{T}} W_u \bar{D}\bar{N} W_e + \Pi \bar{B}\bar{N} W_e\right)\right)^{\mathrm{T}} \end{bmatrix}^{\mathrm{T}} \tag{6.28}$$

where $\bar{M}$ and $\bar{N}$ are defined according to

$$\bar{M} = (\bar{\mathsf{A}}^{\mathrm{T}} + \Pi\bar{\mathsf{B}})^{-1} \quad (6.29)$$

$$\bar{N} = -\left(\bar{D}^{\mathrm{T}} W_u \bar{D} + W_e\right)^{-1}. \quad (6.30)$$

In this case, the matrix Π is the definite-positive stationary solution of the following algebraic Riccati equation:

$$\Pi\bar{\mathsf{A}} + \bar{\mathsf{A}}^{\mathrm{T}}\Pi + \Pi\bar{\mathsf{B}}\Pi + \bar{\mathsf{C}} = 0. \quad (6.31)$$

The matrix $\bar{\mathsf{A}}$, $\bar{\mathsf{B}}$, and $\bar{\mathsf{C}}$ are given by

$$\bar{\mathsf{A}} = \bar{A} + \bar{B}\bar{N}\bar{D}^{\mathrm{T}} W_u \bar{C} \quad (6.32)$$

$$\bar{\mathsf{B}} = \bar{B}\bar{N}\bar{B}^{\mathrm{T}} \quad (6.33)$$

$$\bar{\mathsf{C}} = \bar{C}^{\mathrm{T}} W_u (I + \bar{D}\bar{N}\bar{D}^{\mathrm{T}} W_u)\bar{C} \quad (6.34)$$

where $(\bar{A}, \bar{B}, \bar{C}, \bar{D})$ denotes the state-space matrices of $\overline{K}(s)$.

Remark 6.2 Using this bumpless strategy in Figs. 6.4 and 6.5, we assume that F_s has access to the controller states $\tilde{x}$ or $\bar{x}$. It's a modest assumption since most modern controllers will be realized in software form. Hence, the states are available from computer variables.

6.4 Application to a B747-100/200

6.4.1 Requirements and Validation Tools

The benchmark used for preliminary evaluations has been developed within a GARTEUR project (FM-AG16; see Chap. 1). It corresponds to a highly representative nonlinear aircraft model based on the Boeing 747-100/200. This model originally developed under Matlab/Simulink® environment (see [35, 36]), can accurately simulate real-life conditions and the performance of an aircraft. The Matlab/Simulink® model has been later enhanced in [37–40] for the integrated assessment of GARTEUR's FTC methods. As a part of the FM-AG (16) project, six faulty scenarios have been implemented. Five of them deal with control surface failures, and the last is concerned with the El Al Flight 1862 catastrophic accident [41]. The faulty situation investigated in this paper consists of the motion of the extreme positive position of the Trimmable Horizontal Stabilizer (THS) surface at the maximum rate limit (i.e., $+0.5°/s$). This THS fault is assumed to correspond to a

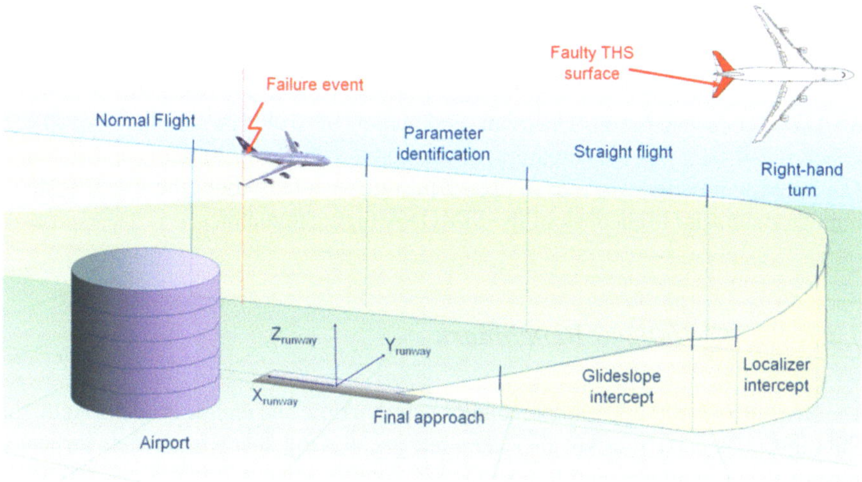

Fig. 6.6 Evaluated test maneuvers in FM-AG(16) project

Fig. 6.7 The SIMONA flight simulator

hardware malfunction. Hence, it is assumed that it is not possible to act on the faulty THS surface to accommodate this fault or put the surface into its neutral position. The flight scenarios (see Fig. 6.6 for a detailed description of the considered flight scenario) have been selected to provide challenging assessment criteria to evaluate the effectiveness and potential benefits of the investigated FTC methods.

For final evaluation, a full flight 6-axes simulator[2] has been used. The flight simulator is presented in Fig. 6.7. It is a six degrees-of-freedom hydraulic motion system

[2]This flight simulator (SIMONA) is located at the Delft University of Technology, Netherlands

tuned to give the pilot realistic inertial motion cues in nominal and failure conditions, specifically developed for human–machine interface and handling qualities research [42–45]. The simulator's flexible architecture [42] allows for a high-fidelity integration of the B747-100/200 Matlab/Simulink® model by using the Real-Time Workshop code generation. The inputs and outputs of SIMONA simulator have been standardized such that the actuators are driven by a dSPACE/Simulink architecture. The interested reader can refer to [42–45] for more details about the visual system, simulator cab, and flight desk of the SIMONA flight simulator.

6.4.2 B747-100/200 Benchmark

The Boeing 747-100/200 model includes aerodynamic and engine models. Actuator and sensor characteristics are also taken into account together with models for wind, atmospheric turbulence, and faults [37, 38]. The aerodynamic forces and moments are defined in terms of aerodynamic coefficients. These coefficients are given in the form of look-up tables. The dimension of the aircraft output vector is 142. However, all output signals are not necessary to control the aircraft. Indeed, the FCS (presented in the next section) uses only 16 measured signals.

The dynamical behavior of the aircraft is described by the following nonlinear state representation

$$\dot{x}_{\mathrm{NL}}(t) = f\left(x_{\mathrm{NL}}(t), u_{\mathrm{NL}}(t), w(t)\right) \tag{6.35}$$

$$y_{\mathrm{NL}}(t) = g\left(x_{\mathrm{NL}}(t), u_{\mathrm{NL}}(t)\right) + v(t) \tag{6.36}$$

where x_{NL}, u_{NL}, and y_{NL} are the state, input, and output vectors, respectively, of the full aircraft nonlinear model. The input and state components are given in Appendix A (see Tables A.1 and A.2). The signal w denotes the process noise related to, e.g., winds and atmospheric turbulences. The signal v represents the measurement noises which are assumed to be uniformly distributed random signals. In this model, physical parameters, e.g., mass, inertia, are fixed to their nominal values. The interested reader can refer to [35] or [5] for a complete description of the aircraft output vector y_{NL}.

Once a trim condition is established for the nonlinear aircraft model, a linear model is generated to capture the dynamics [46]. Simplified models for the longitudinal and lateral modes can then be derived to gain a better physical insight into the modes and their interactions. These models are widely used in aeronautical engineering and are not developed here (see [46] for more details). Since the THS is a symmetric surface, THS faults act mainly on the longitudinal motion (lateral motion effects are neglected). Therefore, the following simplified state-space model derived from (6.35) and (6.36) is retained to describe the dynamics of the aircraft

$$\begin{cases} \dot{x}(t) = A(\rho)x(t) + B(\rho)u(t) + E(\rho)w(t) \\ y(t) = C(\rho)x(t) + v(t) \end{cases} \tag{6.37}$$

where $u = (\delta_{e\bullet\bullet}, i_h)^{\mathrm{T}}$ is the control input vector defined by elevators and THS deflections, respectively. The vector $y = (q, \theta, \dot{h}, h)^{\mathrm{T}}$ is the measured output where q, θ, $\dot{h}$, and h correspond to the pitch rate, pitch angle, altitude rate, and altitude, respectively. The longitudinal state vector is defined by $x = (q, V_{\mathrm{TAS}}, \alpha, \theta, h)^{\mathrm{T}}$, where V_{TAS} and α denote the true airspeed and the angle of attack. The vector of parameters ρ models the varying aerodynamic coefficients. Note that ρ is a function of α (angle of attack) and β (angle of sideslip).

Taking into account the THS faults, the following linear state-space model is derived from (6.37) by assuming that ρ is close to its nominal value during the considered flight trajectory (longitudinal flight):

$$\begin{cases} \dot{x}(t) = Ax(t) + B_e\delta_{e\bullet\bullet}(t) + B_h f_{\mathrm{THS}}(t) + Ew(t) \\ y(t) = Cx(t) + v(t) \end{cases}. \tag{6.38}$$

Here, B_e and B_h are matrices of appropriate dimensions deduced from the B matrix in (6.37). The numerical values of A, B_e, B_h, and C are given in Appendix B. The input signal $\delta_{e\bullet\bullet}$ corresponds to the elevator defections and f_{THS} denotes the THS fault under consideration. Note that this model is clearly an approximation of the real aircraft faulty behavior. However, extensive simulations have shown that this approximation is sufficient for the purpose of the analysis offered here (see [14] for more details).

The goal is the design of an FTC scheme which provides safe accommodation without making any change to the nominal control laws. Before proposing a FTC scheme, it is then required to model the already in-place control system. The SIMONA control architecture is standardized as presented in Fig. 6.8. Motion Control Computer (MCC) corresponds to a standard B747 autoflight system composed by one flight control system (FCS) and a path-planning unit which generates reference trajectories. The MCC inputs are the manual pilot inputs, the Mode Control Panel (MCP) inputs, and the sensor data bus. As mentioned above, the MCC outputs permit to drive the actuators by using a dSPACE/Simulink© architecture. Two control modes are implemented in the simulator: in the manual control mode, the aircraft is only controlled by the FCS, while in the automatic control mode, the FCS is fitted with the path-planning unit.

To model the FCS and because the THS faults act mainly on the longitudinal motion, only the longitudinal part of the FCS is discussed here (see Fig. 6.9 for an illustration). As it can be seen, the elevator and the THS deflections are defined according to the following control laws:

$$\delta_e(t) = K_1(\delta_{\mathrm{col}}, \varepsilon)\delta_{\mathrm{col}}(t) \tag{6.39}$$

$$i_h(t) = K_2(\varepsilon)i_{\mathrm{href}}(t) \tag{6.40}$$

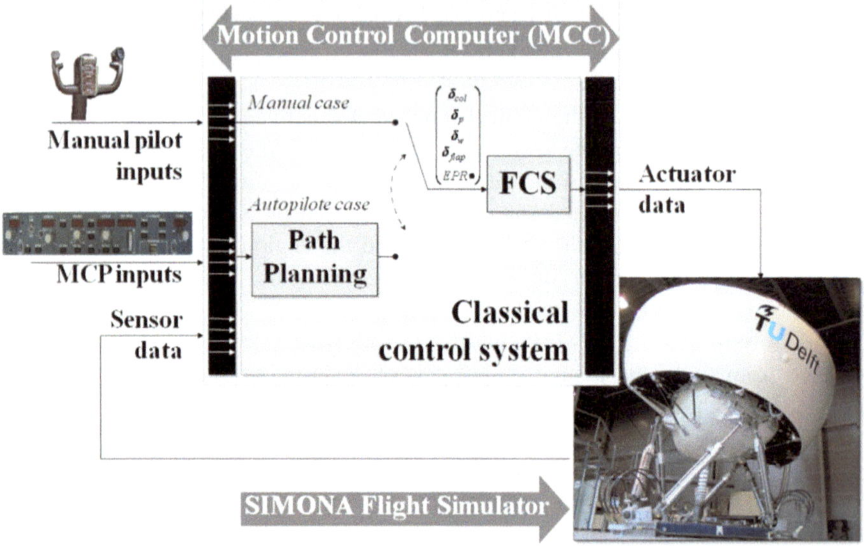

Fig. 6.8 Control law setup

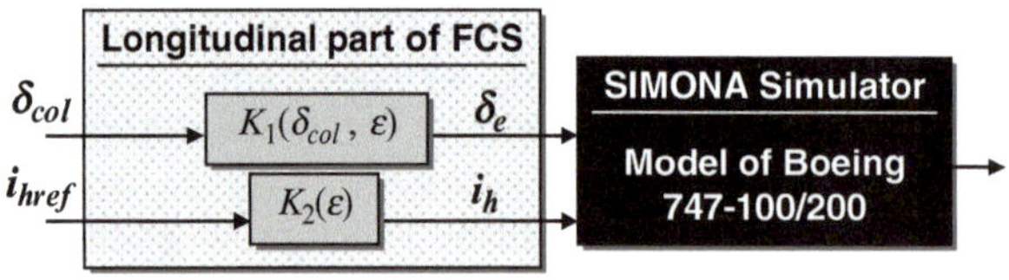

Fig. 6.9 FCS unit for longitudinal motion

where δ_{col} and i_{href} are the reference inputs to elevator and THS surfaces; the parameter ε denotes a vector tuned to achieve flight performances. For instance, ε is a function of the dynamic pressure $\bar{q}$ (the interested reader can refer to [47] for more details). FCS is thus in a gain-schedule-based controller where the scheduling parameters are δ_{col} and ε.

6.4.3 *FTC Problem Formulation to FM-AG(16) Project*

In GARTEUR FM-AG(16) project, an onboard FDI scheme (hardware redundancy) was available. It is assumed that the detection delay does not exceed 500 ms [48]. Hence, the main issue is to show how the FTC mechanism will recover system stability and performance in the worst-case situation, i.e., for the worst time delay (500 ms). The FTC problem is thus formulated by using the materials given in

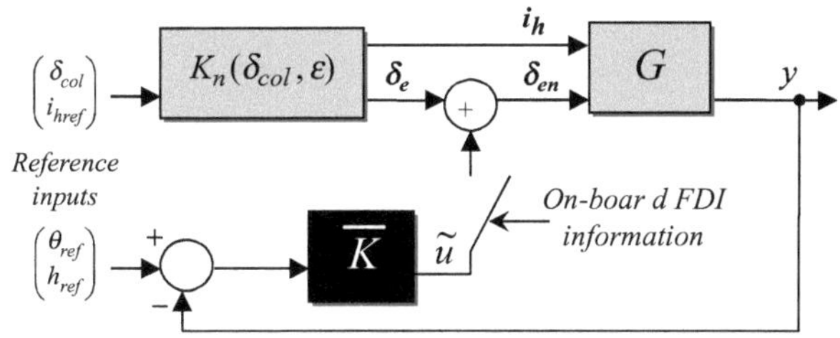

Fig. 6.10 Retained fault-tolerant flight control architecture

Sect. 6.2.2. To deal with manual and autopilot cases, the retained FTC structure is given in Fig. 6.10 where θ_{ref} and h_{ref} are the pitch angle and altitude reference signals provided by the onboard path-planning unit (not developed here; see [47] for more details).

The proposed scheme is composed of three parts:

- $K_n(\delta_{\text{col}}, \varepsilon)$ corresponds to the longitudinal FCS unit (see Eqs. (6.39) and (6.40)).
- G is the model of aircraft dynamics (see Sect. 6.4.2).
- $\overline{K}$ represents a FTC part which generates an additional control signal $\tilde{u}$ to be added to the nominal control signal δ_e computed according to (6.39).

The overall FTC strategy works in such a way that, in a fault-free situation, the FTC loop is not activated leaving the aircraft only controlled by the in-place FCS. When the THS fault is detected by the onboard fault detection unit, the FTC part is just activated via a switching logic.

Based on Problem 6.2, the FTC problem is defined as follows:

Problem 6.3 Assume that a solution to the FTC problem exists, i.e., the effects of faults are compensable (this assumption will be discussed in the next subsection). The goal is to design a controller $\overline{K}(s)$ to produce the new control signal

$$\delta_{\text{en}}(s) = \delta_e(s) + \overline{K}(s)\begin{pmatrix}\theta_{\text{ref}}(s) - \theta_{\text{measured}}(s)\\ h_{\text{ref}}(s) - h_{\text{measured}}(s)\end{pmatrix} \tag{6.41}$$

such that the stability of the feedback system illustrated in Fig. 6.10 and the required control specifications are guaranteed for the considered faulty situation. In an H_∞ setting, this statement means that $\overline{K}(s)$ should satisfy

$$\| F_l(P(s), \overline{K}(s)) \|_\infty < \gamma_2 \tag{6.42}$$

where $F_l(P(s), \overline{K}(s))$ denotes the lower linear fractional transformation of $P(s)$ by $\overline{K}(s)$. The transfer matrix $P(s)$ is deduced from $G(s)$ after some algebraic manipulations (detailed in the next section,) and γ_2 denotes a given performance level to achieve. □

6.4.4 FTC Design

To start, let us answer the following question: Does there exist a "solution" that fully compensates the fault effects by acting on the remaining healthy surfaces? To answer this question, let us look at first the fault compensability problem.

6.4.4.1 Fault Compensability

This problem can be formulated in terms of flight envelope regions in the "altitude-true airspeed" space where the aircraft cannot be rotationally balanced in the presence of faults. This trim deficiency analysis can be formulated according to the following nonlinear constrained optimization problem:

Problem 6.4 Consider the nonlinear model of the aircraft defined by (6.35). The problem consists in finding a combination of unsaturated fault-free control surfaces in the presence of faults that ensures the static equilibrium of the aircraft around its center of gravity during its maneuvers. This problem can be formulated as follows:

$$\min_{x_{\mathrm{NL}},u_{\mathrm{NL}}} \left(W_p \parallel \dot{p} \parallel_2 + W_q \parallel \dot{q} \parallel_2 + W_r \parallel \dot{r} \parallel_2\right) \tag{6.43}$$

$$\text{s.t.} \begin{cases} \left|u_{\mathrm{NL}}^i\right| \le \max(u_{\mathrm{NL}}^i) \\ u_{\mathrm{NL}}^f = \tau \end{cases},$$

$$\forall h \in [h_{\min}, h_{\max}] \text{ and } \forall V_{\mathrm{TAS}} \in [V_{\mathrm{TAS\ min}}, V_{\mathrm{TAS\ max}}]$$

where p, q, and r are the roll, pitch, and yaw rates, respectively. u_{NL}^f is the component of u_{NL} related to the considered fault, and W_p, W_q, and W_r are a priori chosen weighting functions. u_{NL}^i refers to the fault-free actuators, i.e., the remaining healthy control surfaces. $\max(u_{\mathrm{NL}}^i)$ denotes the physical limitations of the ith actuator. τ is a scalar representing the position of the considered faulty surface. □

A point in the "altitude-true airspeed" space is considered "trim deficient" if the solution to the above optimization problem leads to a criterion higher than a prescribed value χ corresponding to the boundary between the trimmable and non-trimmable regions. As a consequence, a fault is considered non-compensable if the flight trajectory of the aircraft (projected in the "altitude-true airspeed" space) crosses a non-trimmable region.

Figure 6.11 depicts the results of this analysis in fault-free (left) and faulty (right) situations during a straight flight at 1,000 m and heading 0° ((*i*) phase), a change of altitude ((*ii*) phase), a right turn ((*iii*) phase), and an altitude change

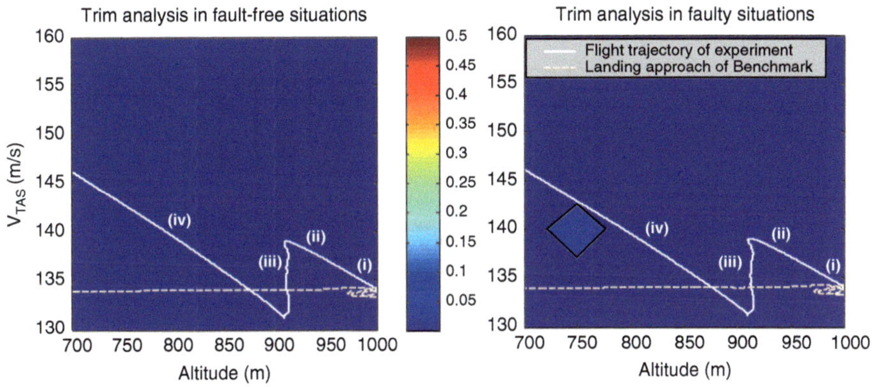

Fig. 6.11 THS fault compensability

((*iv*) phase). This flight corresponds to the one run by the pilot during the experiment on SIMONA flight simulator, as it will be presented in experiment section. The considered fault corresponds to the THS surface jammed on its extreme positive position (+3°). Here, there is no reason to privilege $\dot{p}$ with regards to $\dot{q}$ or $\dot{r}$ in the minimization problem (6.43). Thus, the W_p, W_q, *and* W_r weights have been identically chosen, *i.e.*, $W_p = W_q = W_r = 1$. As it can be seen, all regions are trimmable in fault-free situations, whereas in faulty situations, there exists a weak non-trimmable region. However, for the considered faults, it can be seen that the flight trajectory does not cross these critical regions. This indicates that, from theoretical point of view, the considered fault is fully compensable for all considered maneuvers.

6.4.4.2 FTC Design Formulation

Having shown that the considered fault is compensable, let us consider now the design of the FTC loop. Consider the setup shown in Fig. 6.10 and suppose that a fault has occurred and thus that the FTC loop is activated. Let us denote by $\left(\bar{A}, \bar{B}, \bar{C}, \bar{D}\right)$ the matrices of the state-space model associated with $\overline{K}(s)$. Figure 6.10 leads to the following dynamic equations:

$$\dot{x}(t) = (A - B_e \bar{D} C)x(t) + B_e \bar{C} \bar{x}(t) + B_h f_{\mathrm{THS}}(t) + B_e \delta_e(t) \tag{6.44}$$

$$\dot{\bar{x}}(t) = -\bar{B} C x(t) + \bar{A} \bar{x}(t) \tag{6.45}$$

$$y(t) = C x(t) \tag{6.46}$$

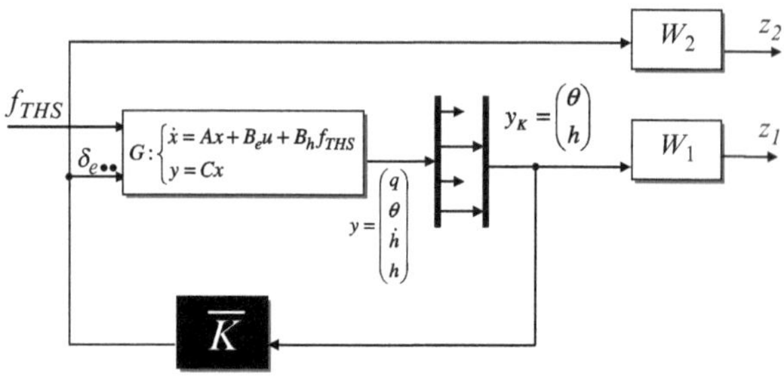

Fig. 6.12 FTC design problem

where x and $\bar{x}$ denote the state vector defined in (6.38) and those associated to $\overline{K}(s)$, respectively. From (6.44), (6.45), and (6.46), the FTC loop state-space model $G_{\text{FTC}}(s)$ which is the transfer between $(\,f_{\text{THS}}\;\delta_e\,)^{\text{T}}$ and y is defined according to

$$G_{\text{FTC}}:\begin{cases}\begin{pmatrix}\dot{x}\\ \dot{\bar{x}}\end{pmatrix} = \begin{pmatrix}A - B_e\bar{D}C & B_e\bar{C}\\ -\bar{B}C & \bar{A}\end{pmatrix}\begin{pmatrix}x\\ \bar{x}\end{pmatrix} + \begin{pmatrix}B_h & B_e\\ 0 & 0\end{pmatrix}\begin{pmatrix}f_{\text{THS}}\\ \delta_e\end{pmatrix}\\ y = (C\;0)\begin{pmatrix}x\\ \bar{x}\end{pmatrix}\end{cases}. \tag{6.47}$$

This expression shows that the stability of the overall loop depends on $(\bar{A}, \bar{B}, \bar{C}, \bar{D})$. As discussed in Proposition 6.1., as long as $\|G_{\text{FTC}}(s)\|_\infty \leq \|G(s)\|_\infty$, stability of the FTC loop and flight performances are preserved, despite the presence of faults. Moreover, $\overline{K}(s)$ must be stable since it operates in open loop in fault-free situations. Hence, Propositions 6.1 and 6.2 are used to formulate the FTC design problem within a H_∞ "mixed-sensitivity" setting fitted to strong stabilization constraint.

6.4.4.3 Design of $\overline{K}(s)$

The goal is now to compute the gain $\overline{K}(s)$ following the aforementioned H_∞ strong stabilization technique. Here, the design objectives will be related to the control law error signal and the control magnitude since the desired performances (no actuator saturation phenomena, tracking the reference trajectory) can be fully achieved. Then, the problem turns out to be the design of a stable controller $\overline{K}(s)$ such that (6.12) and (6.13) are satisfied. The setup used for this design framework is given in Fig. 6.12. $W_1(s)$ and $W_2(s)$ are the weighting functions used to shape the transfer functions $S_{\text{FTC}}(s)$ and $R_{\text{FTC}}(s)$ that are defined according to

$$S_{\text{FTC}}(s) = \left(I + MG_e(s)\overline{K}(s)\right)^{-1} MG_h(s) \tag{6.48}$$

$$R_{\text{FTC}}(s) = \overline{K}(s)S_{\text{FTC}}(s) \tag{6.49}$$

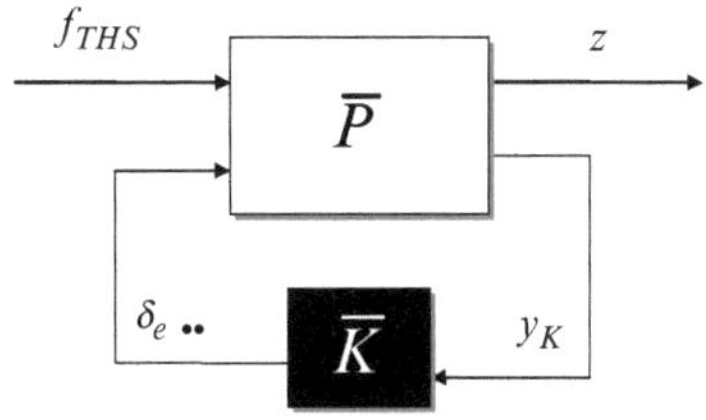

Fig. 6.13 Standard design setup

where the matrix $M = \begin{pmatrix} 0\,1\,0\,0 \\ 0\,0\,0\,1 \end{pmatrix}$ is introduced to select h and θ from y (see Fig. 6.12 for easy reference). The transfer functions $G_e(s)$ and $G_h(s)$ are, respectively, given by

$$G_u(s) = C(sI - A)^{-1}B_e \tag{6.50}$$

$$G_h(s) = C(sI - A)^{-1}B_h \tag{6.51}$$

where A, B_e, B_h, and C matrices are defined according to (6.38).

The weighting function $W_1(s)$ is chosen to achieve small damping ratio on altitude h (m) and pitch angle θ (rad) in a faulty situation. Moreover, an integral component is introduced in $W_1(s)$ to guarantee that the aircraft keeps its trajectory despite the fault. $W_2(s)$ has been fixed to take into account actuator saturation phenomena. More precisely, $W_2{}^{-1}$is a low-pass filter to attenuate the energy of the control signal applied to elevator surfaces such that the control signal behavior keeps "smooth" (high-frequency filter action). The final choices for $W_1(s)$ and $W_2(s)$ are

$$W_1(s) = \left(18\frac{0.5s + 1}{5 \times 10^{-2}s + 1}; 1 \times 10^5 \frac{50s + 1}{1 \times 10^7 s + 1}\right) = (W_\theta(s); W_h(s)) \tag{6.52}$$

$$W_2(s) = \begin{pmatrix} 0.1\frac{0.1s+1}{2.5\times10^{-4}s+1} & 0 & 0 & 0 \\ 0 & 0.1\frac{0.1s+1}{2.5\times10^{-4}s+1} & 0 & 0 \\ 0 & 0 & 0.1\frac{0.1s+1}{2.5\times10^{-4}s+1} & 0 \\ 0 & 0 & 0 & 0.1\frac{0.1s+1}{2.5\times10^{-4}s+1} \end{pmatrix}. \tag{6.53}$$

Using some direct linear fractional algebraic manipulations (LFT), the problem illustrated in Fig. 6.12 is transformed to the problem presented in Fig. 6.13 where $\bar{P}(s)$ is defined according to

$$\begin{pmatrix} z_1(t) \\ z_2(t) \\ y_K(t) \end{pmatrix} = \bar{P}(s)\begin{pmatrix} f_{\text{THS}}(t) \\ \delta_{e\bullet\bullet}(t) \end{pmatrix} \Leftrightarrow \bar{P}(s) = \begin{pmatrix} W_1(s)MG_h(s) & W_1(s)MG_e(s) \\ 0 & W_2(s) \\ MG_h(s) & MG_e(s) \end{pmatrix}. \tag{6.54}$$

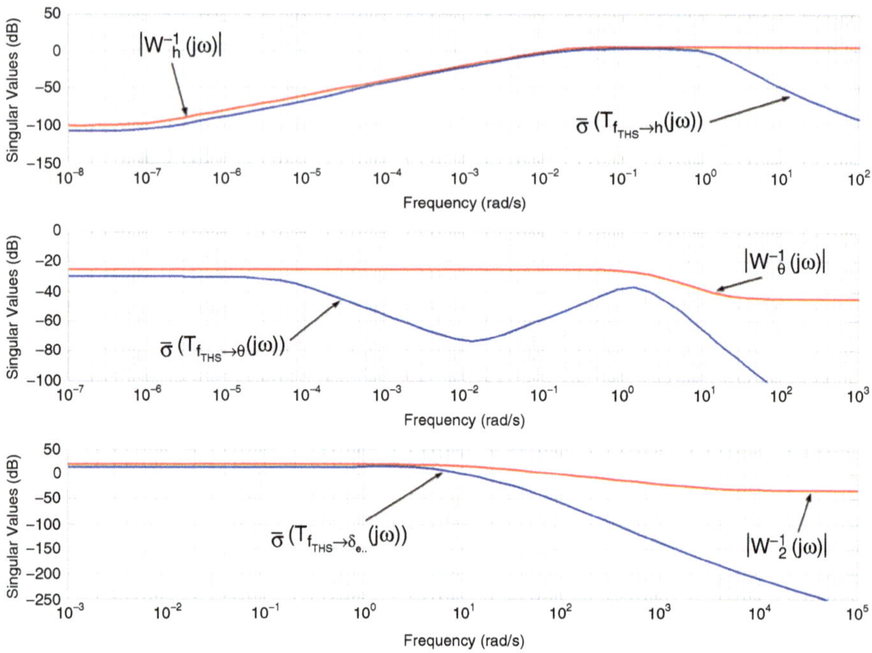

Fig. 6.14 Post-analysis of $\overline{K}(s)$

$\overline{K}(s)$ can now be designed. Here, the central controller, i.e., $\overline{K}(s) = F_l(\hat{K}(s), 0) = \hat{K}(s)$, is retained since $\hat{A}$ is found stable (see Proposition 6.2). Figure 6.14 shows frequency responses obtained for this solution. As it can be seen, the singular value of all sensitivity functions are below the objective weighting functions. This feature indicates that the computed FTC controller $\overline{K}(s)$ achieves the desired performance level. In addition, the weak gaps between the blue and red lines indicate that the nominal performances of the flight control law are preserved.

6.4.5 *Simulation and Experimental Results*

The controller $\overline{K}(s)$ is first implemented within the Matlab/Simulink® simulation benchmark, and after its validation through extensive simulations, within the SIMONA flight simulator. The architecture used to implement $\overline{K}(s)$ in the SIMONA simulator is presented in Fig. 6.15. Figures 6.16, 6.17, and 6.18 present simulation results from Matlab/Simulink® simulator, whereas Figs. 6.19 and 6.20 are devoted to the pilot experiment. Recall that the faulty scenario corresponds to a hardware malfunction of the THS surface (THS surface moves quickly to the extreme position of $+3°$) and cannot be accommodated by the in-place control laws (see [14]).

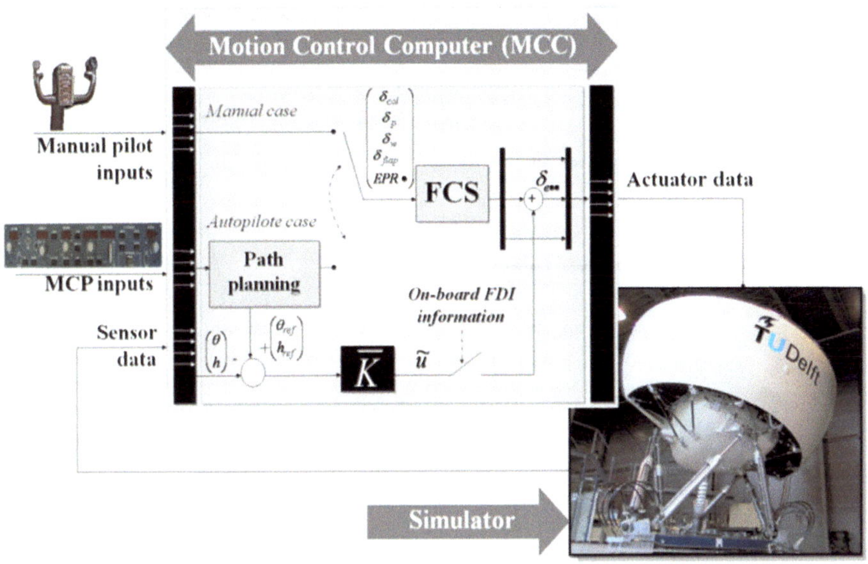

Fig. 6.15 Fault flight tolerant control architecture in SIMONA simulator

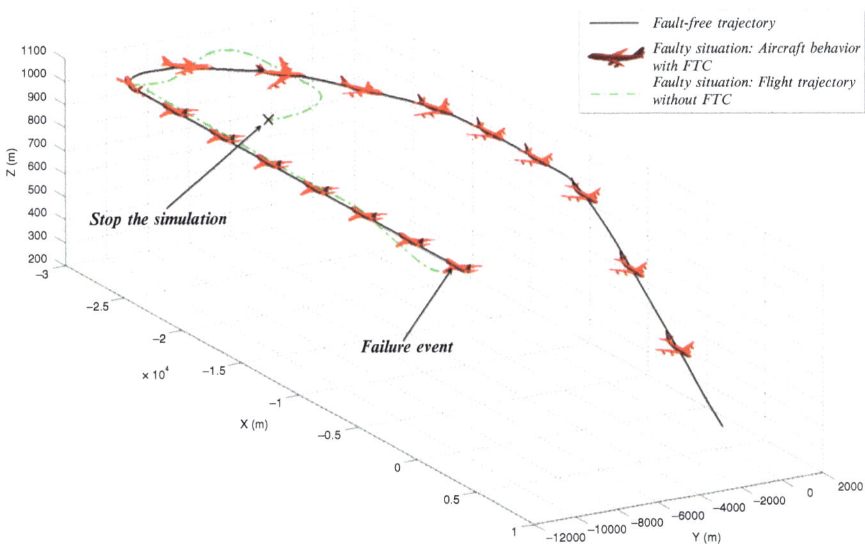

Fig. 6.16 Aircraft response to THS runaway fault

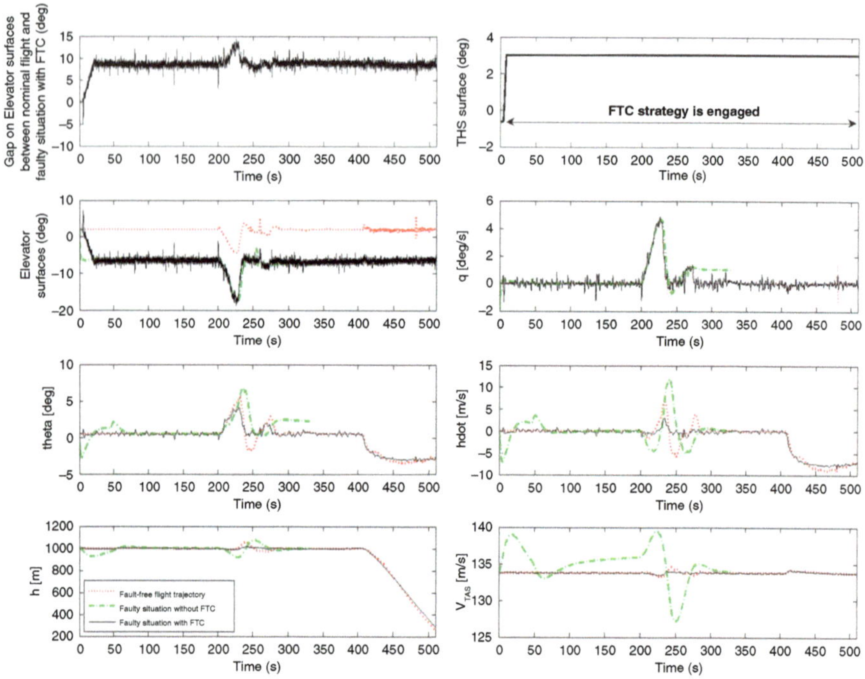

Fig. 6.17 Aircraft response due to THS runaway fault

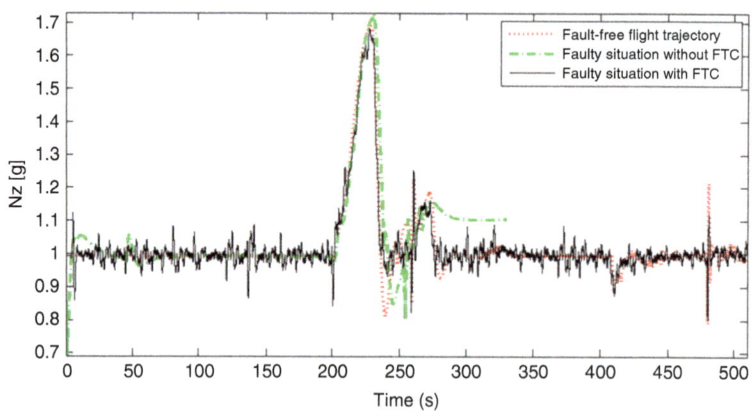

Fig. 6.18 Load factor: THS runaway fault

6.4.5.1 Matlab/Simulink® Benchmark Results

Figure 6.16 presents the nominal fault-free trajectory (landing approach) and highlights the benefit of the proposed scheme through the behavior of the aircraft when the proposed FTC strategy is active. The case where the aircraft is only

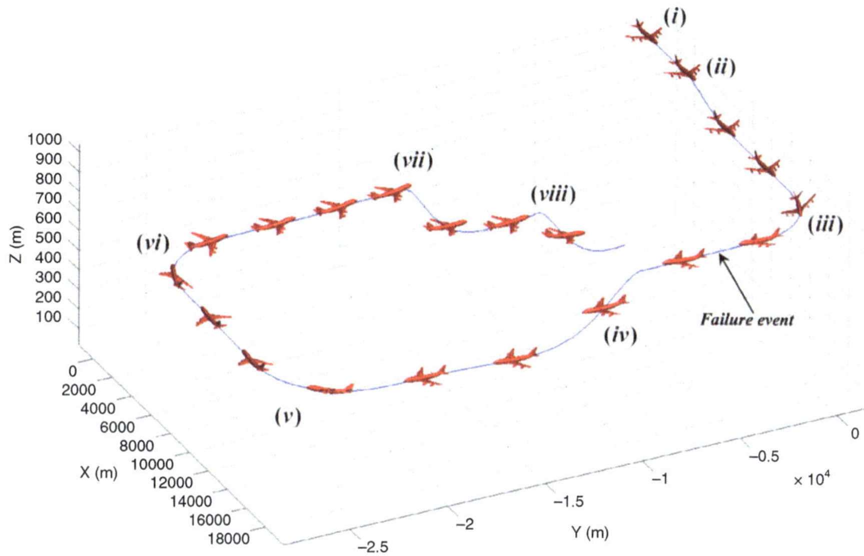

Fig. 6.19 Aircraft flight trajectory: THS runaway fault

controlled by the already in-place control laws is also presented. As it can be seen from this figure in the latter case, the aircraft does not follow the nominal trajectory and lose the trajectory after the first turn. However, when the proposed FTC strategy acts, the aircraft keeps normal flight trajectory, *i.e.*, the aircraft lands successfully despite the fault.

To get a deeper insight into the situation, Fig. 6.17 provides more details about the behavior of the aircraft via the altitude h, the pitch rate q, the velocity V_{TAS}, the pitch angle θ, the altitude rate $\dot{h}$, and the control signals $\delta_{e\bullet\bullet}$. To emphasize the benefit of the proposed FTC scheme, the same simulation is performed when the aircraft is only controlled by the in-place dedicated control systems (no FTC) and when the FTC loop acts. The plots are given for a flight of 510 s. As it can be seen from Fig. 6.17, when the FTC scheme is in place, the controlled faulty system keeps the nominal flight trajectory, i.e., the nominal landing approach. Furthermore, it can be seen that, as expected, the elevator deflections do not exceed the position and rate limits (the deflection and rate limits for the elevators are $[-23°; +17°]$ and $\pm 37°/s$, respectively).

Figure 6.18 illustrates the behavior of the load factor n_z. As it can be seen, the magnitude of undesirable transients on n_z is acceptable since the load factor behavior in fault-free situation is similar to the behavior obtained with the proposed FTC strategy in faulty situation. Note that the pick value of n_z corresponds to the flight situation where the roll angle is near its critical value (55°). It can be seen from Figs. 6.17 and 6.18 that this critical flight phase is well managed by the proposed scheme since there are no saturation phenomena of elevator surfaces and, as previously mentioned, the magnitude of n_z is similar to fault-free case.

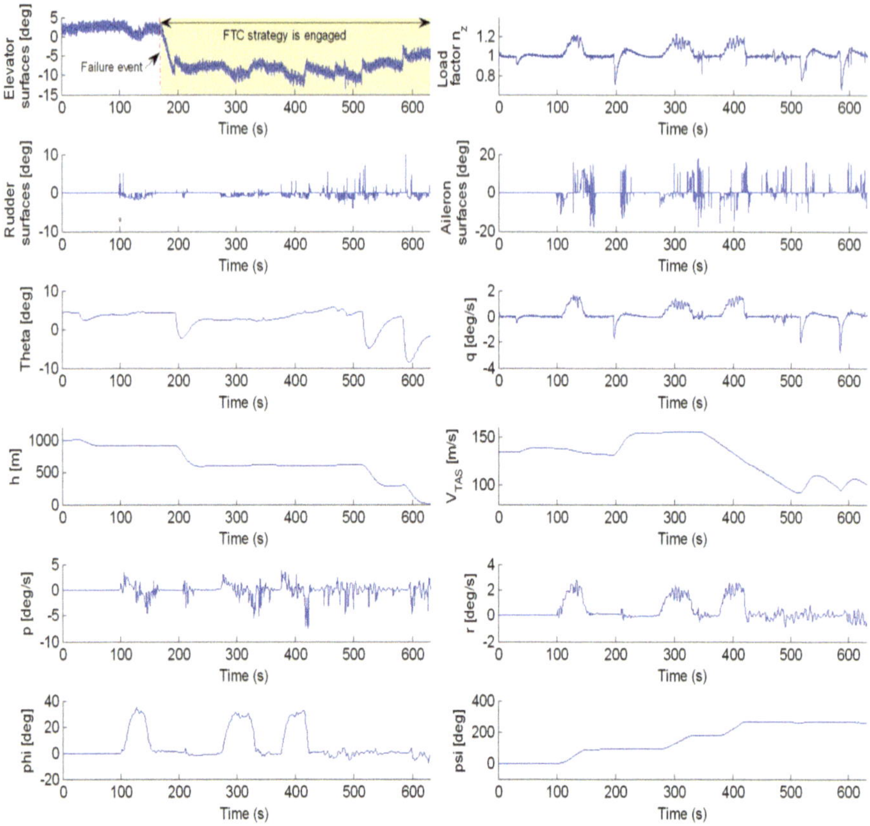

Fig. 6.20 Aircraft responses due to THS runaway fault

6.4.5.2 Experiment Results

The pilot experiment is now considered. The tested scenario is as follows:

(i) The flight starts at 1,000 m, heading 0°.
(ii) First, the altitude reference is changed to 915 m.
(iii) It is followed by a manual heading change to 90°. When the aircraft is stable on the new heading, the stabilizer runaway failure is introduced (at $t = 172$ s).
(iv) An altitude change to 610 m is done (*iv*) using the MCP (see Fig. 6.3 for easy reference of the Mode Control Panel).
(v) It is followed by another manual turn to 180°. Thrust is kept at a minimum.
(vi) When nearing the localizer, a final turn is done to 270° and the localizer signal is manually captured. In the meantime, flaps are being deployed.

Table 6.1 Computation time

Description of the control system	CPU time (ms)
FCS	20
FCS with FTC strategy	28

(vii) On glide slope capture, another altitude change is done to 330 m using the MCP.
(viii) Descending at low speed, we again capture the glide slope and a final altitude change is done to 33 m. The run ends very near the threshold, with correct speed for landing.

Figure 6.19 illustrates the behavior of the aircraft. It can be seen that the aircraft lands successfully despite the presence of the THS fault.

Figure 6.20 gives more information about the behavior of the aircraft. As it can be seen, the FTC strategy ensures a quick compensation of the considered THS fault. Moreover, the tracking of the aircraft altitude is successfully achieved with an almost null damping ratio. Furthermore, it can be seen that this fault is fully accommodated without exceeding the limits of the elevator surfaces (recall that the deflection limits for the elevators are$[-23°; +17°]$). This piloted experiment demonstrates the potential of the proposed FTC scheme.

6.4.5.3 Additional Evaluation Criterion

Finally, the computation time of the proposed FTC strategy is evaluated and compared to the nominal FCS control law, for a possible real-time implementation in a flight control system. This evaluation is done using built-in SIMONA procedures and will not be described here. The results (see Table 6.1) show that it takes about more than 8 ms for the control law to be computed when the FTC scheme is engaged. This computation time has been judged acceptable.

6.5 Conclusion

This chapter presented an active fault-tolerant flight control scheme. The presented techniques have been validated through piloted flight simulator experiments. The faulty situation studied corresponds to the movement of an extreme position of the THS surface during landing approach. A key feature of the proposed approach is that the design of FTC loop is achieved by keeping unchanged the existing flight control system. Once a fault is detected, the control law is, in real time, reconfigured to accommodate the fault. The required additional computational time is low, making the proposed scheme a viable candidate for onboard implementation.

Appendix A: State and Input Definition of the Boeing 747-100/200

Table A.1 State definition of the Boeing 747-100/200

Symbol	Name	Unit
x_{NL} (1): p	Body roll rate	rad/s
x_{NL} (2): q	Body pitch rate	rad/s
x_{NL} (3): r	Body yaw rate	rad/s
x_{NL} (4): V_{TAS}	True air speed	m/s
x_{NL} (5): α	Angle of attack	rad
x_{NL} (6): β	Angle of sideslip	rad
x_{NL} (7): φ	Angle of roll	rad
x_{NL} (8): θ	Angle of pitch	rad
x_{NL} (9): ψ	Angle of yaw	rad
x_{NL} (10): h	Altitude	m
x_{NL} (11): x_e	Distance in X_e-direction	m
x_{NL} (12): y_e	Distance in Y_e-direction	m
x_{NL} $(12+k)$, $k=1, \ldots, 130$	State components related to actuator, sensor, and disturbance models	–

Table A.2 Input definition of the Boeing 747-100/200

Symbol	Name	Unit
u_{NL} (1): $\delta_{a\bullet\bullet}$	4 aileron deflections	deg
u_{NL} (2): $\delta_{sp\bullet}$	12 spoilers	deg
u_{NL} (3): $\delta_{e\bullet\bullet}$	4 elevator deflections	deg
u_{NL} (4): i_h	Stabilizer deflection	deg
u_{NL} (5): $\delta_{r\bullet}$	2 rudder deflections	deg
u_{NL} (6): $\delta_{f\bullet}$	2 flap deflections	deg
u_{NL} (7): $EPR\bullet$	4 thrust engine position	–
u_{NL} (8): $gear$	Gear position	–

Appendix B: A, B_e, B_h, and C State-Space Matrices

$$A = \begin{pmatrix} -6.7926\times10^{-1} & -8.6\times10^{-6} & -8.856\times10^{-1} & 0 & -3.45\times10^{-6} \\ -1.6179\times10^{-1} & -7.588\times10^{-3} & 4.9965 & -9.8 & 4.59\times10^{-5} \\ 1.0084 & -1.0036\times10^{-3} & -6.735\times10^{-1} & 0 & 5.9\times10^{-6} \\ 1 & 0 & 0 & 0 & 0 \\ 0 & 0 & -1.338\times10^{2} & 1.338\times10^{2} & 0 \end{pmatrix}$$

$$B_e = \begin{pmatrix} -4.965\times10^{-3} & -4.965\times10^{-3} & -4.764\times10^{-3} & -4.764\times10^{-3} \\ 0 & 0 & 0 & 0 \\ -1.86\times10^{-4} & -1.86\times10^{-4} & -1.9\times10^{-4} & -1.9\times10^{-4} \\ 0 & 0 & 0 & 0 \\ 0 & 0 & 0 & 0 \end{pmatrix}$$

$$B_h = \begin{pmatrix} -4.5944\times10^{-2} \\ 0 \\ -1.912\times10^{-3} \\ 0 \\ 0 \end{pmatrix}$$

$$C = \begin{pmatrix} 1 & 0 & 0 & 0 & 0 \\ 0 & 0 & 0 & 1 & 0 \\ 0 & 0 & -1.338\times10^{2} & 1.338\times10^{2} & 0 \\ 0 & 0 & 0 & 0 & 1 \end{pmatrix}$$

References

1. Patton R (1997) Fault tolerant control: the 1997 situation. In: 3th symposium on fault detection, supervision and safety for technical processes (SAFEPROCESS'97). Hull University, Hull, UK, pp 1033–1054
2. Blanke M, Kinnaert M, Lunze M, Staroswiecki M (2006) Diagnosis and fault tolerant control. Springer, New York
3. Noura H, Theilliol D, Ponsart JC, Chamseddine A (2009) Fault-tolerant control systems: design and practical applications. Springer, London
4. Zhang Y, Jiang J (2008) Bibliographical review on reconfigurable fault-tolerant control systems. Annu Rev Control 32:229–252
5. Edwards C, Lombaerts T, Smaili H (2010) Fault tolerant flight control: a benchmark challenge, Lecture notes in control and information sciences. Springer, Berlin
6. Staroswiecki M (2008) On fault handling in control systems. Int J Control Autom Syst Spec Issue FDI FTC 6(3):1–10
7. Cieslak J, Henry D, Zolghadri A (2010) Fault tolerant flight control: from theory to piloted flight simulator experiments. IET Control Theory Appl 4(8):1451–1461
8. Staroswiecki M, Yang H, Jiang B (2007) Progressive accommodation of parametric faults in LQ control. Automatica 43:2070–2076
9. Hamdaoui R, Ponsart JC, Theilliol D (2009) Study of an actuator fault accommodation based on LQ control energy criterion. In: 7th workshop on advanced control and diagnosis
10. Lombaerts T, Chu P, Mulder JA, Joosten D (2009) Flight control reconfiguration based on a modular approach. SAFEPROCESS'2009. Barcelona, Spain, pp 259–264
11. Maciejowski J, Jones C (2003) MPC fault-tolerant flight control case study: Flight 1862. SAFEPROCESS'2003, pp 121–126
12. Joosten DA, van den Boom TJJ, Lombaerts TJJ (2007) Effective control allocation in fault-tolerant flight control with MPC and feedback linearization. In: Proceedings of the European conference on systems and control. Kos, Greece, July 2007

13. Zhou K, Ren Z (2001) A new controller architecture for high performance, robust, and fault-tolerant control. IEEE Trans Autom Control 46:1613–1617
14. Cieslak J, Henry D, Zolghadri A, Goupil P (2008) Development of an active fault tolerant flight control strategy. AIAA J Guid Control Dyn 31(1):135–147
15. Gaspar P, Szaszi I, Bokor J (2005) Reconfigurable control structure to prevent the rollover of heavy vehicles. Control Eng Pract 13:699–711
16. Cui P, Weng Z, Patton R (2008) Novel active fault-tolerant control scheme and its application to a double inverted pendulum system. J Syst Eng Electron 19(1):134–140
17. Alwi H, Edwards C, Stroosma O, Mulder JA (2008) Fault tolerant sliding mode control design with piloted simulator evaluation. AIAA J Guid Control Dyn 31(5):1186–1201
18. Alwi H, Edwards C (2008) Fault tolerant control using sliding modes with on-line control allocation. Automatica 44:1859–1866
19. Alwi H, Edwards C, Tan CP (2011) Fault detection and fault-tolerant control using sliding modes, Advances in industrial control. Springer, London
20. Shin J-Y, Belcastro CM (2006) Performance analysis on fault tolerant control system. IEEE Trans Control Syst Technol 14(9):1283–1294
21. Doyle JC, Glover K, Khargonekar P, Francis B (1989) State-space solutions to standard H_2 and H_∞ control problems. IEEE Trans Autom Control 34(8):831–847
22. Gahinet P, Apkarian P (1994) A linear matrix inequality approach to H_∞ control. Int J Nonlinear Control 4:421–428
23. Yang H, Jiang B, Staroswiecki M (2009) Supervisory fault tolerant control for a class of uncertain nonlinear systems. Automatica 45:2319–2324
24. Yang H, Jiang B, Cocquempot V (2010) Fault tolerant control design for hybrid system, Lecture notes in control and information sciences. Springer, Berlin
25. Jain T, Yamé JJ, Sauter D (2011) Model-free reconfiguration mechanism for fault tolerance. Int J Appl Math Comput Sci 22(1):125–137
26. Efimov D, Cieslak J, Henry D (2010) Supervisory fault tolerant control via common Lyapunov function approach. In: IEEE conference on control and fault tolerant systems
27. Efimov DV, Cieslak J, Henry D (2012) Supervisory fault tolerant control with mutual performance optimization. Int J Adapt Control Signal Process. doi: 10.1002/acs.2296
28. Yamé JJ, Hanping Q, Kinnaert M (2010) A self-conditioned implementation of switching controllers for smooth transition in multimode systems. In: IEEE conference on control and fault tolerant systems
29. Cieslak J, Efimov D, Henry D (2012) Supervisory fault tolerant control scheme based on bumpless scheme and dwell-time conditions. IFAC Safeprocess'12, Mexico City, Mexico
30. Cieslak J, Efimov D, Henry D (2010) Observer-based structures to Active fault tolerant control problems. In: IEEE conference on control and fault tolerant systems
31. Cieslak J, Henry D, Zolghadri A (2010) A solution for management of fault diagnostic and fault tolerance performances in active FTC schemes. In: 18th IFAC symposium on automatic control in aerospace, Nara, Japan
32. Zhou K, Doyle JC (1997) Essentials of robust control. Prentice Hall, Upper Saddle River
33. Campos-Delgado DU, Zhou K (2003) A parametric optimization approach to H_∞ and H_2 strong stabilization. Automatica 39:1205–1211
34. Turner MC, Walker DJ (2000) Linear quadratic bumpless transfer. Automatica 36:1089–1101
35. Van der linden, CAAM (1996) DASMAT-Delft university aircraft simulation model and analysis tool. Report LR-781, Technical University Delft, Delft
36. Smaili MH (1999) FLIGHTLAB 747 benchmark for advanced flight control engineering v4.03. Technical report, Technical University Delft, Delft
37. Marcos A, Balas G (2003) A Boeing 747-100/200 aircraft fault tolerant and diagnostic benchmark. Department of Aerospace and Engineering Mechanics, TR AEM-UoM-2003-1, University of Minnesota, Minneapolis, MN
38. Smaili H, Breeman J, Lombaerts T, Stroosma O (2009) A benchmark for fault tolerant flight control evaluation. SAFEPROCESS'2009, Barcelona, Spain

39. Smaili MH, Breeman J, Lombaerts TJJ, Joosten DA (2006) A simulation benchmark for integrated fault tolerant flight control evaluation. In: AIAA modeling and simulation technologies conference and exhibit, 21–24 Aug 2006, Keystone, CO
40. Lombaerts TJJ, Joosten DA, Breeman J, Smaili H, Van den Boom TJJ, Chuk QP, Mulder JA, Verhaegen M (2006) Assessment criteria as specifications for reconfiguring control. In: AIAA guidance, navigation, and control conference and exhibit, 21–24 Aug 2006, Keystone, CO
41. Smaili MH, Mulder JA (2000) Flight data reconstruction and simulation of the 1992 Amsterdam Bijlmermeer airplane accident. AIAA GNC conference. AIAA 2000–4586
42. Koekebakker SH (2001) Model based control of a flight simulator motion base. Ph.D. thesis, Delft University of Technology, Delft
43. Stroosma O, Van Paassen MM, Mulder M (2003) Using the SIMONA research simulator for human-machine interaction research. In: AIAA modeling and simulation technologies conference. Austin, TX, Aug 2003
44. Berkouwer WR, Stroosma O, Van Paassen MM, Mulder M, Mulder JA (2005) Measuring the performance of the SIMONA research simulator's motion system. In: AIAA modeling and simulation conference. San Francisco, CA, Aug 2005
45. Stroosma O, Smaili H, Mulder JA (2009) Pilot-in-the-loop evaluation of fault-tolerant flight control systems. SAFEPROCESS'2009. Barcelona, Spain, pp 259–264
46. Boiffier JL (1998) The dynamics of flight equations. John Wiley & Sons, Chichester UK
47. Hanke C, Nordwall DR (1970) The simulation of a jumbo jet transport aircraft Volume II: modeling data. Boeing Company, D6-30643
48. Varga A (2007) Fault detection and Isolation of actuator failures for a large transport aircraft. In: European air and space conference. Germany

Chapter 7
Model-Based FDIR for Space Applications

Acronyms

ACC	ACCelerometer
AOCS	Attitude and Orbit Control System
CSS	Coarse Sun Sensor
FEEP	Field Emission Electric Propulsion
GNSS	Global Navigation Satellite System
GYR	GYRoscope
IMU	Inertial Measurement Unit
LFR	Linear Fractional Representation
LIDAR	LIght Detection And Ranging
LMI	Linear Matrix Inequality
MAV	Mars Ascent Vehicle
MDV	Mars Descent Vehicle
MSR	Mars Sample Return
NAC	Narrow Acquisition Camera
NAV	NAVigation
NEP	Nominal Exit Point
PID	Proportional Integral Derivative
RFS	Radio Frequency Sensor
RLV	Reentry Launch Vehicle
RW	Reaction Wheel
SAM	Sun Acquisition Mode
SDP	Semi-Definite Programming
STR	Star TRacker
TAEM	Terminal Area Energy Management
TEP	TAEM Exit Point
THR	THuRster
TM	Telemetry

A. Zolghadri et al., *Fault Diagnosis and Fault-Tolerant Control and Guidance for Aerospace Vehicles: From Theory to Application*, Advances in Industrial Control,
DOI 10.1007/978-1-4471-5313-9_7,

7.1 Introduction

This chapter is dedicated to actuator fault detection and diagnosis in space applications. Fault tolerance in terms of control and guidance will also be discussed. The design method is based on H_∞/H_- and robust pole assignment tools. Three space applications will be studied:

- The first one deals with a satellite example, namely, Microscope.[1] A model-based strategy is presented for detecting and isolating faults which can occur in the satellite thruster actuation system. The existing GNC system for Microscope will be described. The aim is to diagnose failures that correspond to thrusters blocking themselves and/or closing when in operation, despite the presence of measurement noises, delays, sensor misalignment phenomena, and spatial disturbances (i.e., third-body disturbances, J2 disturbances, atmospheric drag, and solar radiation pressure).
- The second space application deals with a deep space mission, the so-called Mars Sample Return (MSR) mission.[2] The objective is to develop a fault diagnosis scheme to detect and isolate faults occurring in the orbiter thrusters, despite the presence of unknown but bounded delays induced by the electronic devices and the uncertainties on the thruster rise times.
- The third application is an atmospheric reentry vehicle.[3] The goal is to detect and isolate any kind of faults in the wing flap actuators during the auto-landing and "Terminal Area Energy Management" (TAEM) phase. A key feature of the proposed approach is that the coupling between the in-plane and out-of-plane vehicle motions as well as the effects that faults could have on the guidance, navigation, and control performance are explicitly taken into account within the design procedure.

7.2 FDIR in Space Applications: State-of-Practice

Satellites and spacecraft have considerably evolved over the last decades from preprogrammed automata performing a priori known tasks and unable to react against unforeseen events to smart embedded system able to take preprogrammed decision on event occurrence or able to react against context changes. Onboard FDIR solutions have become now integral elements in designing health monitoring systems for space systems.

[1]This case study is taken from a collaborative project supported by the French Space Agency (CNES).

[2]The Mars Sample Return (MSR) mission is taken from a European project supported by the European Space Agency and Thales Alenia Space (France).

[3]HL-20 vehicle: This study is taken from a collaborative project (SICVER project, see Chap. 1) with European Space Agency and EADS Astrium Space Transportation (France).

Reaction time for detection and fault isolation and robustness and ability to recover from a failure are sizing elements of the satellite/spacecraft availability. For instance, for missions requiring a medium level of availability such as scientific Earth observation mission where short mission interruptions may be allowed, satellite saving will be preferred than mission follow-on. When a fault is detected, the satellite is fully reconfigured and the failed unit is switched to a redundant one. This is sometimes called a "half-satellite" strategy [1, 2]. The ground stations are in charge of planning appropriate corrective actions. This strategy is very basic and requires a weak validation effort. It is applicable only for mission requiring a small availability rate, and it guarantees to keep the satellite in a safe mode. On the other hand, when availability level should be high, for instance, in telecommunication satellites or deep space missions (such as Cassini–Huygens, Exomars, Mars Sample Return, Mars Express), FDIR schemes are often more complex and structured in different levels in order to avoid that the effect of a fault leads to a reduction of mission outage. When a fault occurs, the satellite/spacecraft should be kept in an operational mode. For such cases, the strategy used is based on a set of hierarchical levels enabling for a graduated reaction. Such a hierarchical structure helps recover the fault with a quick reaction time and minimize the perimeter of their effects. Each failure is recovered at the lowest layer to limit the impact on the mission.

Generally speaking, for space missions, the fault management architecture is composed by four levels which have different reaction times and are activated successively by order of criticality. The faults are filtered in each level so that a higher level can only be called under specific conditions, for example, when the lower level has been activated several times. The higher the level is, the more critical is the fault and the lower is the probability occurrence of the faults. In terms of validation, the cost and the effort are very high due to the complexity of this architecture. Nevertheless, the main advantage of such architecture is the possibility to activate or deactivate one or several levels depending on the mission requirements. Figure 7.1 illustrates this hierarchical FDI(R) strategy:

- Level 0 deals with failures having no impact on the satellite/spacecraft subsystem performance and matches faults which can be recovered by local correction (bit flip, cycle redundancy check . . .). Detection is performed internally in the units. The recovery is autonomous and local to the unit.
- Level 1 deals with failures requiring switching a unit to its redundant one. Detection is performed outside the unit and the recovery is done by the subsystem in which the unit is involved. The effect of such a failure can lead to a temporary degraded mode without any effect on the mission goals. By instance, when a sensor fails, the subsystem can use the last measure to perform its processing or extrapolate the next values in some cases.
- Levels 2 and 3 are often mixed due to the fact they have the same kind of detection and recovery action. They deal with loss of performance for a subsystem. Level 2 is strictly related to the occurrence of several alarms coming from lower levels. This means that the recovery actions which have been engaged have not corrected the anomaly and that the fault has to be considered more globally at subsystem

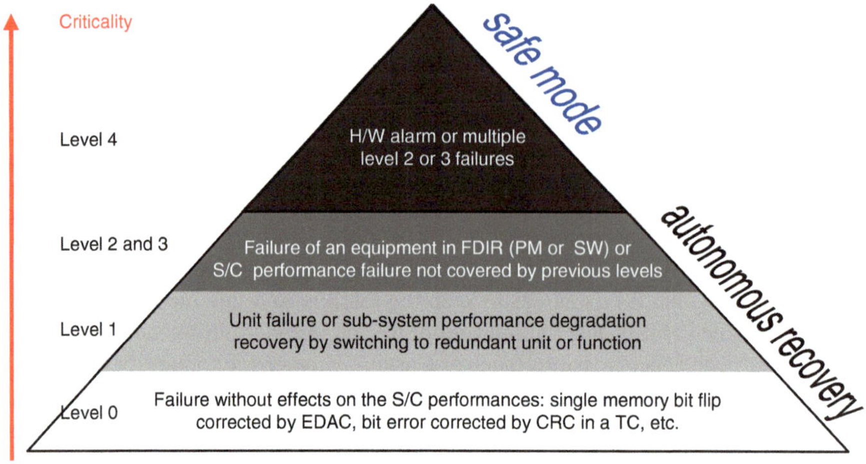

Fig. 7.1 Today's fault diagnosis and management architecture for satellite/spacecraft

or platform level. Level 3 is related to faults on the FDIR units like software or processor module. These faults are recovered by switching to a redundant processor module. Level 2/3 recovery is still considered autonomous for most of these faults even if some main functions can be impacted.

- The most critical level is the level 4, which is activated in case of several alarms from level 2/3 or from hardware alarms. This kind of alarm is the last one which can be raised by the satellite and are due to a critical breakdown leading to the safe mode. The recovery in this case has to be done by the ground stations and the mission is interrupted.

Through these four levels the identification is seldom considered, due to the fact that the detection is judged sufficiently reliable to perform the identification in the same time. Most of the detection and recovery actions are software except for level 4. Degraded modes are associated to the faults activating level 1 or 2. In case of level 2 occurrences, mission performance is often degraded leading to a loss of performance for the accomplishment of mission goals, but the mission follows on.

Detection at level 1 is generally based on checks: internal unit checks, data transmission checks, and consistency checks. Internal checks are the simplest approach to unit/component detection. When the unit allows, it performs some internal monitoring and provides a health status. This status is reported in telemetry (TM) for the ground stations and used onboard to raise local alarm. Data transmission checks aim at detecting protocol communication anomalies. It concerns all the units connected to a data communication network. Protocol is often based on a secured and real-time one, which enables the access to a set of indicators describing the data transmission status. Check-sum is used to monitor the data validity. All these elements are part of the protocol and are part of the FDIR information used to detect misbehavior of an embedded data communication network. Consistency checks are used to complete the two previous kinds of detections and to address the monitoring of the data's value. The two previous detection solutions deal only with unit health status and data transmission. It concerns again all the onboard sensors and actuators.

Model-based FDI techniques can be envisaged as complementary solutions of those being used in level 1 within the global fault management architecture described above [1–3].

7.3 Model-Based FDIR Solutions

There are plenty of model-based diagnostic techniques that have been considered for potential application to space systems. See, among others, the references reported in Chap. 2. The precursor technique is probably the static parity space method applied for fault diagnosis in the inertial measurement units (IMUs) [4–6]. The approach relies on the redundant measurements acquired from the IMUs for deriving the so-called parity space relations. In particular, two configurations were considered, namely, the octahedron configuration and the dodecahedron configuration and, more recently, the dedicated pyramidal configuration [2]. Some other studies are based on particle-filtering-based algorithms (see, for instance, [7–9]). Basically, a particle filter is a Markov chain Monte Carlo algorithm that performs system state estimation using a set of samples (particles). In [10] the proposed method is based on an FDI observer combined with a residual weighting matrix. The proposed design procedure involves eigenstructure assignment. In [11] directional nonlinear observers are used to detect and isolate faults in small satellite actuators. The idea is to design the observer gain so that the kth component of the residual changes in a definite direction if and only if a fault occurs in the kth actuator. Other FDI techniques are based on the so-called sliding mode observers [12–14]. The approach consists in performing an estimate of the fault rather than detecting the presence of it through a residual signal. The method could provide also a direct estimate of the fault's size. In [15] robust dynamic observers, organized as a bank of estimators, are used to generate the residual signals. Selected performance criteria indices are also used, together with Monte Carlo robustness tuning and performance evaluation, to provide fault diagnosis solutions. The particle filtering approach has been considered by [16, 17] for a K9-planetary rover and the Hyperion experiment. The unknown input observer technique was considered for thruster fault detection of the Mars Express spacecraft in [15, 18, 19]. The sliding mode observer (SMO) approach was considered for the Mars Express experiment in [20]. The fault detection and isolation problem was concerned by the thrusters and the (four) gyro system during the Sun Acquisition Mode (SAM). Norm-based approaches were applied to actuator faults in the shuttle orbiter during the transonic regime in [21]. The norm-based approach was also applied to actuator faults detection problem of the Hopper reentry vehicle in [22], and control surface FDI is considered in the HL-20 reentry vehicle during the landing phase in [23, 24]. The norm-based approach was also considered for micro-Newton colloidal thruster faults during the experiment phase for the LISA Pathfinder experiment in [25, 26], and finally, thruster and gyro faults during station keeping maneuvers were considered for telecom satellites. See, for example, many NASA technical reports[4] available for other specific case studies.

In this chapter and among a large number of possible model-based solutions, we focus on model-based FDI methods that use $H_\infty, H_-, H(0)$ criterion with robust pole assignment techniques. The approach provides several attractive characteristics. Firstly, it offers tunable design parameters through the so-called weighting functions that allow the designer to specify the fault detection performance/robustness specifications and trade-offs. Secondly, it offers a reasonable computational burden since the final fault diagnosis scheme results in simple LTI (linear time invariant) filters. The technique corresponds to a complete design/analysis cycle. As it will be seen, the method provides a practically relevant and general framework in which various design goals and trade-offs are formulated and managed. The design problem can be solved by numerically powerful LMI-based techniques. Moreover, a systematic analysis procedure allows the designer to check if all FDI objectives are achieved given the model uncertainties and disturbances. This problem results in a min–max optimization problem that can be formulated and solved using a *generalized μ-analysis* procedure. The degree of conservativeness of the FDI design can then be quantified through this post-design analysis, allowing the user to get a clear idea on how the design trade-offs should be retuned to get as close as possible to the required FDI performance levels. The methodological foundations are mainly underpinned by [27–29].

7.4 Notations

The following notations will be used throughout the chapter: R and C denote the real and complex sets, respectively. $A^T\,(A^*)$, $A > 0$ means the transpose (conjugate) and the definite positiveness of A, respectively. $A \geq (>) B$ means $A - B$ is positive semi-definite positive definite). $\bar{\sigma}(A)/\underline{\sigma}(A)$ denotes the maximum/minimum singular values of the matrix A. $||w||_2$ is used to denote the L_2-norm of the signal w. $P(s)$, or simply P, is assumed to be in $RH_\propto$, real rational function with ($||P||_\infty$ is also the largest gain of P)

$$||P||_\infty = \sup_\omega \bar{\sigma}(P(j\omega)) < \infty. \tag{7.1}$$

In accordance with the induced norm, the smallest gain of $P(s)$ is defined according to $\inf_\omega \underline{\sigma}\,(P(j\omega))$. It can be verified that for some P, e.g., strict proper P, $\inf_\omega \underline{\sigma}\,(P(j\omega)) = 0$ since the frequency range of interest is infinite. This motivates the

[4] http://www.sti.nasa.gov/

introduction of the nonzero smallest gain of P, i.e., the H_- index, as the restriction of $\inf_{\omega} \underline{\sigma}\,(P(j\omega))$ to a finite frequency domain ω, i.e.,

$$||P||_- = \inf_{\omega \in \Omega} \underline{\sigma}\,(P(j\omega)) < \infty. \tag{7.2}$$

However, it is not a norm since it does not verify the Schwartz inequality [30]. This H_- index is used as a criteria for fault sensitivity performance. Note that H_- index is sometimes called by H_- norm, even if it is not a norm (see, for instance, [31]).

As a direct extension, the H (0) gain is defined according to

$$||P||_0 = \lim_{\omega \to 0} \underline{\sigma}(P(j\omega)) \neq 0 \tag{7.3}$$

which is known as the zero frequency gain (dc-gain). This criterion can be useful to enforce fault sensitivity in terms of static gain.

The notation $\left[\begin{array}{c|c} A & B \\ \hline C & D \end{array}\right]$ refers to the of state–space realization a transfer $G(s) = C(sI - A)^{-1}\ B\ +\ D$. Linear fractional representation (LFR) is extensively used in the chapter. For appropriately dimensioned transfers $N(s)$ and $M(s) = \begin{pmatrix} M_{11}(s) & M_{12}(s) \\ M_{21}(s) & M_{22}(s) \end{pmatrix}$, the lower LFR is defined according to F_l $(M,\ N) = M_{11}(s) + M_{12}\ (s)N(s)(I - M_{22}\ (s)N(s))^{-1}\ M_{21}\ (s)$ and the upper LFR according to $F_u\ (M, N) = M_{22}(s) + M_{21}\ (s)N(s)(I - M_{11}\ (s)\ N(s))^{-1}\ M_{12}\ (s)$, under the assumption that the involved inverses exist. This assumption is discussed when it is judged necessary. Otherwise, it is assumed to be satisfied. For more details on LFR algebra, the interested reader can refer to [32–36].

7.5 A Satellite Example

This section deals with the development of a model-based FDI scheme for a microsatellite called MICROSCOPE.[5]

7.5.1 *Description*

MICROSCOPE is due to be launched in 2016 on a circular and sun synchronous polar orbit at 700 km with ascending and descending nodes at 06:00 and 18:00,

[5]MICROSCOPE is the French acronym of MICRO-Satellite à traînée Compensée pour l'Observation du Principe d'Equivalence.

respectively, with in-flight operation during 1 year. The objectives of the MICROSCOPE experiment is to test of the Einstein's equivalence principle with an accuracy of 100 times better than the one obtained with experiments realized on Earth. To carry out its mission, MICROSCOPE combines two rotation motions: the first one is a rotation around the Earth and the second one is a spin rotation. To control its trajectory, MICROSCOPE uses the coupling of six ultrasensitive accelerometer sensors, a stellar sensor, and a very precise electric propulsion system composed by twelve field-emission electric propulsion (FEEP) thrusters.[6]

If an FEEP thruster fault occurs, the satellite may not compensate for nongravitational disturbance, necessary for its mission. Such faulty situations can, of course, be diagnosed by operators using telemetry information collecting by ground stations. However, the risk of lack of communication between the satellite and the ground stations could lead to significant delays that can lead the mission of MICROSCOPE to be aborted. An onboard model-based solution is developed to detect and isolate faults for early recovery actions. The basic concepts have been reported in [27, 28]. The procedure aims to generate a structured residual vector r in the following general form

$$r(s) = M_y y(s) + M_u u(s) - F(s)\begin{pmatrix} y(s) \\ u(s) \end{pmatrix}, \quad u(s) = K(s)y(s) \qquad (7.4)$$

where r, y, u denote the residual, the measurements, and the control signals, respectively. K refers to the AOCS (attitude and orbit control system), M_y and M_u are the two residuals constant structuring (or allocation) matrices, and $F(s)$ is a (stable) dynamical filter. The proposed method consists in jointly designing M_y, M_u, and $F(s)$ such that the effects that faults have on r are maximized in the H_--norm sense while minimizing the influence of spatial disturbances, in the H_∞ norm sense.

7.5.1.1 Modeling

Figure 7.2 shows the general setup of MICROSCOPE. Star trackers and three-axis accelerometers permit to measure the attitude $\Theta(t)$, the inertial rotational acceleration $\dot{\varpi}(t)$, and the linear acceleration $\Gamma(t)$.

The navigation unit is composed of a so-called hybridation filter that computes estimates $\hat{\Theta}(t)$ and $\hat{\Gamma}(t)$, removing misalignment phenomena and some noises. $\hat{\Theta}(t)$ and $\hat{\Gamma}(t)$ will be used later for the design of the FDI unit. We assume that the navigation unit is not perfect, and thus that there still exists time delays and noises on $\hat{\Theta}(t)$, $\hat{\dot{\varpi}}(t)$ and $\hat{\Gamma}(t)$.

The MICROSCOPE actuation system is composed of 12 FEEP thrusters dispatched at its angles. This enables to control the satellite motion. The open rate

[6]The MICROSCOPE project Steering Committee has authorized the project to start a new preliminary conception phase (phase B) with cold gas micro-thrusters instead of FEEPs.

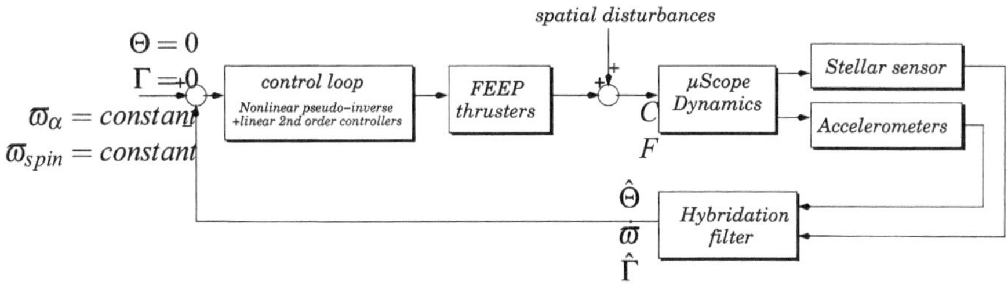

Fig. 7.2 General setup of MICROSCOPE AOCS

of each FEEP thruster is controlled by the correction loop in order to maintain the attitude and the linear acceleration to zero and the orbit rotational velocity ϖ_α and the spin rotational velocity ϖ_{spin} to constant values.

The equations for the rotational motion of MICROSCOPE in the body-fixed axis system (the center of this frame is fixed to the center of mass of MICROSCOPE) are derived from the moment vector equation

$$C = I_s \dot{\varpi} + \varpi \times I_s \varpi \tag{7.5}$$

where "×" denotes the cross product of vectors. I_s is the inertia matrix. C is the moments about the center of mass due to the propulsion and disturbances. $\varpi = (pqr)^{\mathrm{T}}$ is the inertial rotational velocity and $\dot{\varpi}$ is the inertial rotational acceleration. Taking into account the spin rotational velocity ϖ_{spin} of MICROSCOPE, the relation between the rotational velocities and the attitude (Cardan) angles $\Theta = \left(\theta_x \theta_y \theta_z\right)^{\mathrm{T}}$ is given by [37, 38]

$$\begin{aligned} \begin{pmatrix} p \\ q \\ r \end{pmatrix} &= \begin{pmatrix} 1 & 0 & -\sin\theta_y \\ 0 & \cos\theta_x & \sin\theta_x \cos\theta_y \\ 0 & -\sin\theta_x & \cos\theta_x \cos\theta_y \end{pmatrix} \begin{pmatrix} \dot{\theta}_x \\ \dot{\theta}_y \\ \dot{\theta}_z \end{pmatrix} \\ &\quad - \varpi_{\text{spin}} \begin{pmatrix} \cos\theta_y \sin\theta_z \\ \cos\theta_x \cos\theta_y + \sin\theta_x \sin\theta_y \sin\theta_z \\ -\sin\theta_x \cos\theta_z + \cos\theta_x \sin\theta_y \sin\theta_z \end{pmatrix}. \end{aligned} \tag{7.6}$$

The indices x, y, z are referred to the x-, y-, z-axes of the body-fixed frame. For MICROSCOPE, the gravitational forces are compensated by the Coriolis forces. Thus, the satellite linear acceleration that describes the translational motion is given by the following equation:

$$m\Gamma = F + m g_L. \tag{7.7}$$

m denotes the mass of MICROSCOPE. F is the forces due to the propulsion and disturbances. $\Gamma = \left(\Gamma_x \Gamma_y \Gamma_z\right)^{\mathrm{T}}$ is the linear acceleration about the center of mass. g_L denotes the local gravitational field.

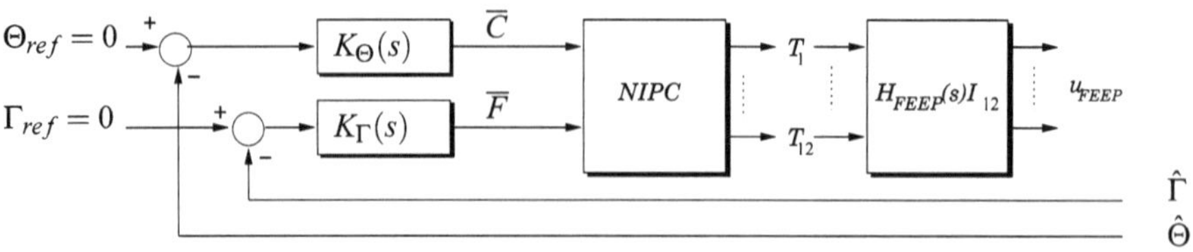

Fig. 7.3 The AOCS of MICROSCOPE

7.5.1.2 Navigation Unit and Actuation System

As mentioned above, the navigation unit is not considered to deliver perfect measurements. Here, we assume that there exist time delays and noises. The numerical values of the time delays have been determined to be 0.1 s for $\hat{\varpi}(t)$ and $\hat{\Gamma}(t)$ measurements and 0.5 s for $\hat{\Theta}(t)$. For both $\hat{\varpi}(t)$ and $\hat{\Gamma}(t)$ measurements, noises are modeled as coloring signals, i.e., they are considered to be filtered Gaussian white noise. These filters are calculated to be of order 6 for the x component and of order 2 for both the y and z components. For $\hat{\Theta}(t)$, we assume simply Gaussian white noises on each axis.

The model describing the dynamics of each FEEP thruster is chosen to be a simply first-order transfer $H_{\text{FEEP}}(s)$ with cutting frequency 2 rad/s, i.e.,

$$H_{\text{FEEP}}(s) = \frac{1}{1+0.5s}. \tag{7.8}$$

7.5.1.3 Control Loops

The control law consists in two second-order linear controllers and a control allocator called NIPC (Nonlinear Iterative Pseudo-inverse Controller). Figure 7.3 illustrates the control law and the NIPC. The control law compensates disturbances which, again, are indispensable prior conditions for testing the equivalence principle.

The controller $K_\Theta(s)$ allows the attitude $\Theta(t)$ to be maintained to zero (i.e., $\theta_x = 0, \theta_y = 0$ and $\theta_z = 0$), and the controller $K_\Gamma(s)$ has been designed in order to maintain the linear acceleration $\Gamma(t)$ to zero (i.e., $\Gamma_x = 0, \Gamma_y = 0$ and $\Gamma_z = 0$). The NIPC allocator manages the open rate of each of the 12 FEEP thrusters. Basically, the NIPC consists in the computation of a matrix inverse: Let T_i be the open rate of the ith thruster. Then, the moments C and the forces F generated by the FEEP thrusters are given by

$$\begin{pmatrix} \bar{C} \\ \bar{F} \end{pmatrix} = M \begin{pmatrix} T_1 \\ \vdots \\ T_{12} \end{pmatrix} \tag{7.9}$$

where $M \in \mathrm{R}^{6\times 12}$ is the thruster configuration matrix. The elements of M define how each thruster affects each component of C and F. Thus, the computation of each T_i, $i = 1, \ldots, 12$ can be done using a simple inversion of the Eq. (7.9). Since M has more columns than rows, there exist an infinite number of solutions. By minimizing a specified criterion, the solution can be made unique, e.g., minimum-power NIPC results in the Moore–Penrose matrix computation M^+. The interested reader can refer to [39, 40] for more details.

7.5.1.4 Disturbances

The disturbances that affect the satellite motion are considered to be due to four phenomena: magnetic, gravitational, aerodynamic, and solar. All phenomena manifest themselves by moments and forces that affect the motion of the satellite that depend on the satellite spin velocity ϖ_{spin} and the orbit velocity ϖ_α. In the subsequent developments, these disturbances will be denoted as $h(\varpi_\alpha, \varpi_{\text{spin}})$. The notation "$(\varpi_\alpha, \varpi_{\text{spin}})$" is used to keep in mind that the disturbances depend on $\varpi_\alpha : \alpha = \varpi_\alpha t$ and $\varpi_{\text{spin}} : \theta_{\text{spin}} = \varpi_{\text{spin}} t$. The interested reader can refer to [41] for a complete mathematical description of $h(\varpi_\alpha, \varpi_{\text{spin}})$.

7.5.1.5 LTI Model of MICROSCOPE

From (7.5), (7.6), (7.7), (7.8), and (7.9) and taking into account the control law, a nonlinear model that describes the overall MICROSCOPE system can be computed. In fact, since the hybridation filter is designed to be robust to the local gravitational field (i.e., the term "g_L" in Eq. (7.7)), it can be verified that Eqs. (7.5), (7.6), (7.7), (7.8), and (7.9) lead to the following model (see Figs. 7.2 and 7.3 for easy reference):

$$\begin{cases} \dot{x} = f(x, \varpi_{\text{spin}}) + E_1 h(\varpi_\alpha, \varpi_{\text{spin}}) + \mathrm{BMC}_P x_P \\ \dot{x}_P = A_P x_P + B_P T \\ \bar{y} = g(x) + E_2 h(\varpi_\alpha, \varpi_{\text{spin}}) + \mathrm{DMC}_P x_P \end{cases} \tag{7.10}$$

$$y_i = e^{-\rho_i s} \bar{y}_i + n_i \tag{7.11}$$

$$T = M^+ \begin{pmatrix} K_\Theta(s)\hat{\Theta} \\ K_\Gamma(s)\hat{\Gamma} \end{pmatrix}. \tag{7.12}$$

The subscript i is used to denote the ith component of a vector. ρ_i denotes the time delay of the ith measurement coming from the navigation unit, i.e., $\rho_i \in \{0.1; 0.5\}s$ (see Sect. 7.5.1.2). $T = (T_1, \ldots, T_{12})^T$ is the controlled input vector due to the propulsion and $y = (\hat{\Theta}^{\mathrm{T}} \hat{\omega}^{\mathrm{T}} \hat{\Gamma}^{\mathrm{T}})^{\mathrm{T}}$ is the measurements vector. n also denotes the associated noises coming from the imperfect navigation unit. $x = (pqr\theta_x\theta_y\theta_z)^{\mathrm{T}}$ is the state vector, and $g(x)$ and $f(x, \varpi_{\text{spin}})$ are nonlinear functions depending on x and the spin rotational velocity ϖ_{spin}. B, D, E_1, and E_2 are constant matrices of appropriate dimensions. A_P, B_P, C_P, and x_P are, respectively, the state matrices and the state vector associated with the transfer $H_{\text{FEEP}}(s)$.

With regard to the faults, we are interested in the FEEP thrusters blocking or closing themselves when in operation. Such faults can be modeled in a multiplicative manner according to (see Chap. 2 and [42] for a discussion about fault classification)

$$u^{f}_{\text{FEEP}}(t) = (I_{12} - \psi)u_{\text{FEEP}}(t), \quad \psi = \text{diag}\{\psi_1, \psi_2, \ldots, \psi_{12}\} \tag{7.13}$$

where $\psi_i, i = 1, \ldots, 12$ are unknown. I_{12} denotes the identity matrix of dimension 12, and u_{FEEP} is the thrust signal applied to the satellite (see Fig. 7.3). The index "f" is used to denote the faulty case. Note that $\psi_i = 1$ indicates that the ith FEEP thruster is closing itself. The case of the ith thruster blocking itself when in operation corresponds to $\psi_i(t) = 1 - \frac{c}{u_{\text{FEEP}_i}(t)}$ where c is a constant value.

Next, substituting C_p in (7.10) by $(I_{12} - \psi)$ and using an approximation of the fault model to get an additive fault description, it follows from (7.10), (7.11), and (7.12) that

$$\begin{cases} \dot{x} = f(x, \varpi_{\text{spin}}) + E_1 h(\varpi_\alpha, \varpi_{\text{spin}}) + \text{BMC}_P x_P + \sum_{i=1}^{12} K_{1_i} f_i \\ \dot{x}_P = A_P x_P + B_P T \\ \bar{y} = g(x) + E_2 h(\varpi_\alpha, \varpi_{\text{spin}}) + \text{DMC}_P x_P + \sum_{i=1}^{12} K_{2_i} f_i \end{cases} \tag{7.14}$$

$$y_i = e^{-\rho_i s}\bar{y}_i + n_i \tag{7.15}$$

$$T = M^{+}\begin{pmatrix} K_\Theta(s)\hat{\Theta} \\ K_\Gamma(s)\hat{\Gamma} \end{pmatrix} \tag{7.16}$$

where (K_{1i}, K_{2i}) is the ith fault signature associated to the ith fault mode f_i. This approximation makes sense as long as the MICROSCOPE control law keeps stability in faulty situations. The interested reader can refer to [43] for a discussion of such an approximation.

Finally, having in mind that MICROSCOPE is controlled around the equilibrium point $\Theta = 0, \Gamma = 0, \varpi_\alpha = \text{constant}$ and $\varpi_{\text{spin}} = \text{constant}$, one can derive from (7.14), (7.15), and (7.16) a linear model by means of a first-order approximation of the nonlinear equations around the equilibrium point. For the time delays ρ_i, a first-order Padé approximation is used. This boils down to the linear time invariant model (LTI)

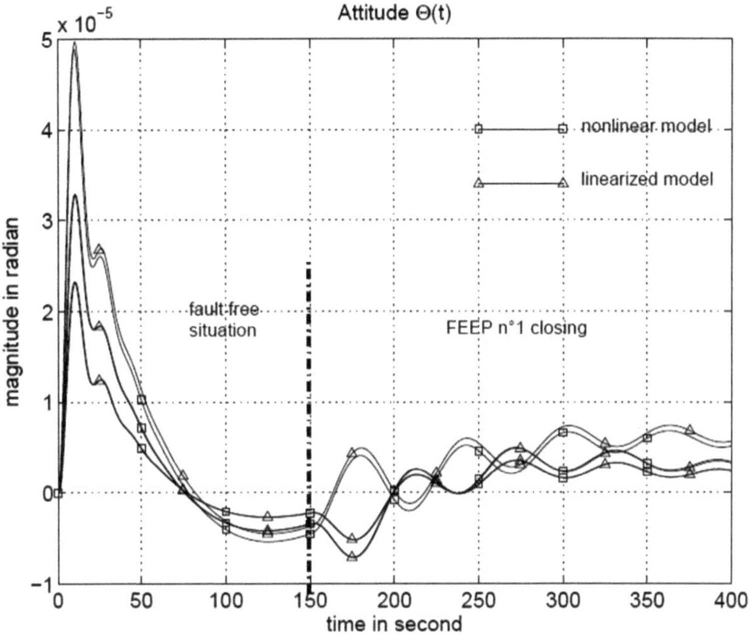

Fig. 7.4 Behavior of $\hat{\Theta}(t)$. Linearized model (*triangle*) versus the nonlinear model (*square*)

$$y = P(s)\begin{pmatrix} f_i \\ h(\varpi_\alpha, \varpi_{\text{spin}}) \\ u \end{pmatrix} + n \tag{7.17}$$

$$u = M^+ \begin{pmatrix} K_\Theta(s)\hat{\Theta} \\ K_\Gamma(s)\hat{\Gamma} \end{pmatrix}. \tag{7.18}$$

To validate the LTI model (7.17) and (7.18), linear simulations were performed versus nonlinear ones. The simulation scenario corresponds to a constant disturbance applied at $t = 0$ s. A fault is simulated in the first FEEP thruster at $t = 150$ s. The goal is to validate the transient behavior and the steady state of the output signals $\hat{\Theta}(t)$, $\hat{\varpi}(t)$, and $\hat{\Gamma}(t)$ predicted by the linearized model in both fault-free and faulty situations. To get a better comparison, the simulations are run without the measurement noises. Figure 7.4 illustrates the behavior of $\hat{\Theta}(t)$ predicted by both the linearized model (plots with triangles) and the nonlinear model (plots with squares). For brevity, the plots of $\hat{\varpi}(t)$ and $\hat{\Gamma}(t)$ are not presented. As it can be seen, the linear model (7.17) and (7.18) approximates well the nonlinear model.

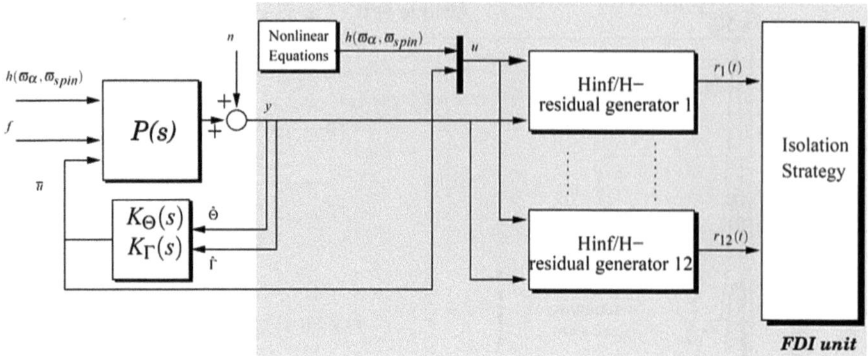

Fig. 7.5 FDI system

7.5.2 Design of the FDI System

In this section, an FDI scheme for detecting and isolating thruster faults despite the presence of the disturbances $h(\varpi_\alpha, \varpi_{\text{spin}})$ and the noises n are considered within the H_∞/H_- setting. The FDI scheme consists of a bank of 12 residual generators that are designed so that the sensitivity level of the ith residual with respect to the ith FEEP thruster fault f_i is maximized in the H_--norm sense while guaranteeing robustness against n and $h(\varpi_\alpha, \varpi_{\text{spin}})$ in the H_∞-norm sense. An original aspect in the proposed scheme is that the a priori knowledge of $h(\varpi_\alpha, \varpi_{\text{spin}})$ is used to manage the 12 residual generators. This enables to enhance the robustness level of the FDI scheme against $h(\varpi_\alpha, \varpi_{\text{spin}})$. Next, the residuals are post-processed by a fault isolation stage in order to isolate the fault uniquely. The technique is based on the evaluation of a cross-correlation criterion between the residuals and the signature that a given FEEP fault has on the controlled thrusters open rate T_i, $i = 1, \ldots, 12$.

The proposed FDI scheme is illustrated on Fig. 7.5. Note that we use the controlled moments and forces $\bar{u} = \left(\bar{C}^{\mathrm{T}} \bar{F}^{\mathrm{T}}\right)^{\mathrm{T}} = \begin{pmatrix} K_\Theta(s)\hat{\Theta} \\ K_\Gamma(s)\hat{\Gamma} \end{pmatrix}$ for the residual generators rather than the controlled open rates T_i, $= 1, \ldots, 12$ of the thrusters.

Let us consider the problem of the design of the ith H_∞/H_- residual generator (this problem is illustrated on Fig. 7.6a). Let the ith residual signal r_i be given by

$$r_i = z_i - \hat{z}_i \tag{7.19}$$

where $\hat{z}_i$ is an estimation of $z_i = M_{y_i} y + M_{u_i} u$, a subset of measurements y and inputs u. $M_{y_i} \in R^{1\times 9}$ and $M_{u_i} \in R^{1\times 12}$ are the two (constant) residual allocation matrices. For clarity, the subscript "i" is ignored from now.

Now, define the augmented disturbance vector as $d = \left(h^{\mathrm{T}}(\varpi_\alpha, \varpi_{\text{spin}})\, n^{\mathrm{T}}\right)^{\mathrm{T}}$ and let the fault f_i be observable from the output y (this assumption is a prior condition for the fault detection problem to be well posed). Then, following the method proposed in [27, 28], the problem turns out to be the design of M_y, M_u, and

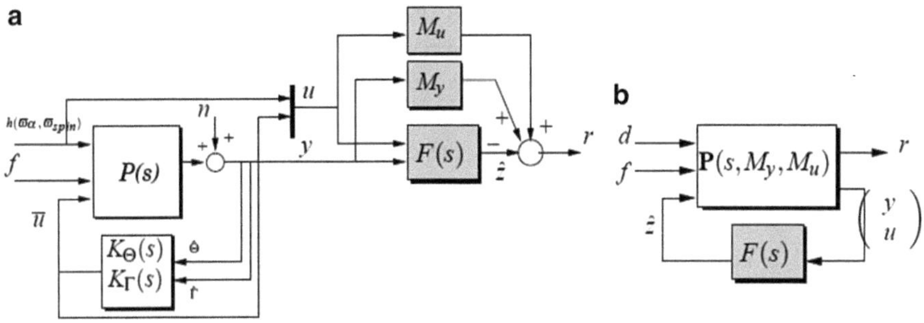

Fig. 7.6 H_∞/H_- fault detector design problem

$F(s) : \hat{z} = F(s) \begin{pmatrix} y \\ u \end{pmatrix}$ such that (see the notation section for definition of the norms and see Fig. 7.6 for easy reference)

- (S.1): $||T_{d \to r}||_\infty < \gamma_1$ where $T_d \to r$ denotes the closed-loop transfer between r and d.
- (S.2): $||T_{f \to r}||_- > \gamma_2$ over a specified frequency range Ω. $T_f \to_r$ denotes the closed-loop transfer between r and f, and Ω is the frequency range where it is required that (S.2) yields.

γ_1 and γ_2 are specified design parameters. In this formulation

$$u = \bar{u} = \left(\bar{C}^{\mathrm{T}} \bar{F}^{\mathrm{T}}\right)^{\mathrm{T}} = \begin{pmatrix} K_\Theta(s)\hat{\Theta} \\ K_\Gamma(s)\hat{\Gamma} \end{pmatrix}, \quad y = \begin{pmatrix} \hat{\Theta}^{\mathrm{T}} \\ \dot{\omega}^{\mathrm{T}} \\ \hat{\Gamma} \end{pmatrix}. \tag{7.20}$$

Note that the problem defined by requirements (S.1) and (S.2) could also be interpreted as a multi-objective optimization problem whereby the choice of γ_1 and γ_2 is guided by the Pareto optimal points. See [44] for a discussion on Pareto multi-criteria optimization problem. However, in practice, γ_1 and γ_2 are better considered as parameters to be tuned by the designer since finding "optimal" values is highly related to the system under consideration (see, for instance, [23, 24, 27, 28, 41, 45]). Here, (S.1) represents the worst-case robustness of the residual to spatial disturbances $h(\varpi_\alpha, \varpi_{\text{spin}})$ and noises n, and (S.2) is the fault sensitivity objective to guarantee high detection performance level. Of course, the smaller γ_1 and the bigger γ_2 are, the better the fault detection performance will be.

The method consists now in formulating the requirements (S.1) and (S.2) in terms of loop shapes, i.e., gain responses for the appropriate closed-loop transfers. These objectives are then stated as uniform bounds by means of the shaping filters.

To proceed, let W_d and W_f denote the (dynamical) shaping filters associated with (S.1) and (S.2), respectively. Due to the definition of d, it is natural to choose W_d according to $W_d = \text{diag}\ (W_h, W_n)$. W_h represents the robustness requirements against the disturbances $h(\varpi_\alpha, \varpi_{\text{spin}})$, and W_n represents the robustness objectives against n. Following the mathematical expression of $h(\varpi_\alpha, \varpi_{\text{spin}})$, it appears that the spatial disturbances appear over the frequency ranges ω_α and $\omega_\alpha - \omega_{\text{spin}}$. Then, it is desired to have a rejecting behavior of $T_{h(\varpi_\alpha,\varpi_{\text{spin}})\to r}(j\omega)$ at those frequencies. $T_{h(\varpi_\alpha,\varpi_{\text{spin}})\to r}(j\omega)$ denotes the components of $T_{d\to r}(j\omega)$ associated with $h(\varpi_\alpha, \varpi_{\text{spin}})$. This leads to choose W_h as a band-stop filter centered to $\omega_c = \frac{2\omega_\alpha - \omega_{\text{spin}}}{2}$ with side-band ξ chosen to cover the frequency range $[\omega_\alpha - \omega_{\text{spin}}; \omega_\alpha]$rd/s, i.e.,

$$W_h = \gamma_h \frac{1 + \frac{2\xi}{\omega_c}s + \frac{1}{\omega_c^2}s^2}{\frac{2\xi}{\omega_c}s(1 + \tau_h s)} I_6. \tag{7.21}$$

The positive constant γ_h is introduced to manage the gain of W_h. This enables to manage the robustness level of the fault detection scheme against the spatial disturbances $h(\varpi_\alpha, \varpi_{\text{spin}})$. The parameter τ_h is introduced to make W_h invertible.

With regard to W_n, since $y = (\hat{\Theta}^{\mathrm{T}} \hat{\dot{\omega}}^{\mathrm{T}} \hat{\Gamma}^{\mathrm{T}})^{\mathrm{T}}$, it is natural to fix W_n according to $W_n = \operatorname{diag}(W_{n\Theta} I_3, W_{n\dot{\omega}} I_3, W_{n\Gamma} I_3)$ where $W_{n\Theta}, W_{n\dot{\omega}}, W_{n\Gamma}$ refer to $\hat{\Theta}, \hat{\dot{\omega}}, \hat{\Gamma}$, respectively. Coming back to the discussion on modeling the navigation unit, it is natural to define $W_{n\dot{\omega}}^{-1}$ and $W_{n\Gamma}^{-1}$ equal to the coloring filters used to model n. As a consequence, the components of $W_{n\dot{\omega}}^{-1}$ and $W_{n\Gamma}^{-1}$ refered to the "x-axes" are high-pass filters of order 6 and the refereed to the "y- and z-axes" are high-pass filters of second order. In fact, in order to reduce the computation time and the order of the fault detection filters, $W_{n\dot{\omega}}^{-1}$ and $W_{n\Gamma}^{-1}$ are chosen equal and as an upper bound of the coloring filters, so that they involve less poles/zeros. This leads to the following definition for

$$W_{n\dot{\omega}} = W_{n\Gamma} = \gamma_{\text{acc}} \frac{(1 + \tau_1 s)^2}{(1 + \tau_2 s)^2} \quad \tau_1 = 0.2s, \tau_2 = 10s. \tag{7.22}$$

The positive constant parameter γ_{acc} is introduced to manage the gain of $W_{n\dot{\omega}} = W_{n\Gamma}$. Clearly, $W_{n\dot{\omega}} = W_{n\Gamma}$ are low-pass filters (remember that $W_{n\dot{\omega}}^{-1}$ and $W_{n\Gamma}^{-1}$ are high-pass filters). In other words, a high-frequency attenuation of $T_{n_{\dot{\varpi}\Gamma}\to r}(j\omega)$ is required at frequencies where the energy content of n is likely to be concentrated, i.e., $\omega \geq 5$rad/s. Here, $T_{n_{\dot{\varpi}\Gamma}\to r}(j\omega)$ denotes the components of $T_{d\to r}(j\omega)$ refereed to $\hat{\dot{\varpi}}$ and $\hat{\Gamma}$. Similarly, $W_{n\Theta}$ is chosen as a constant γ_{att} since for $\hat{\Theta}$, we assumed a white noise.

For the purpose of the fault sensitivity objectives, we consider that the faults appear in low frequencies. This boils down to a first-order low-pass filter for W_f with cutting frequency ω_f, i.e.,

$$W_f = \gamma_2 \frac{1}{1 + \tau_f s} \quad \tau_f = \frac{1}{\omega_f}. \tag{7.23}$$

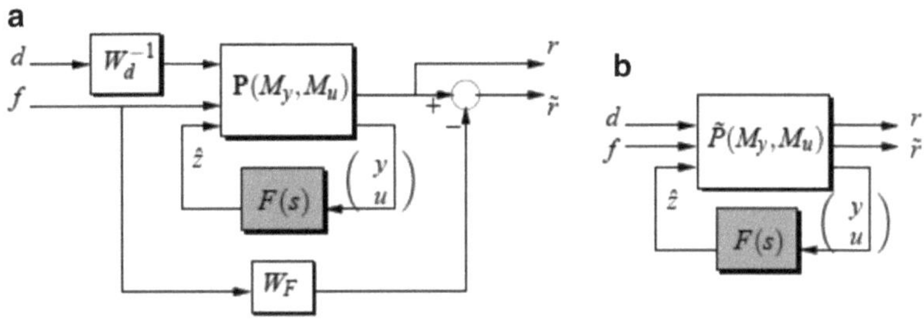

Fig. 7.7 The equivalent forms of the filter design problem

The solution is then handled using the following lemma that basically states that the H_- specification can be solved by solving a fictitious H_∞ problem:

Lemma 7.1 *Consider the robust fault sensitivity specification* (S.2) *and the shaping filter W_f defined by* (7.23). *Introduce W_F, a right invertible transfer matrix, so that $||W_f||_- = \frac{\gamma_2}{\lambda}||W_F||_-$ and $||W_F||_- > \lambda$ where $\lambda = 1 + \gamma_2$. Define the signal $\tilde{r}$ such that $\tilde{r} = r - W_F f$. Then a sufficient condition for the specification* (S.2) *to hold is*

$$||T_{f\to r} - W_F||_\infty < 1 \Leftrightarrow ||T_{f\to \tilde{r}}||_\infty < 1. \tag{7.24}$$

A proof of this lemma is given in [27].

Following Lemma 7.1 and noticing that a necessary and sufficient condition for the robustness requirement (S.1) to hold is

$$||T_{d\to r} W_d^{-1}||_\infty < 1. \tag{7.25}$$

The design problem can be now represented according to the setup depicted on Fig. 7.7a, where $\tilde{d}$ and $\tilde{r}$ are two signals, so that $\tilde{d} = W_d d$ and $\tilde{r} = r - W_F f$. Then including W_d^{-1} and W_F into $\mathbf{P}(M_y, M_u)$ leads to the equivalent block diagram of Fig. 7.7b, where the transfer matrix is deduced from W_d^{-1}, W_F, and $\mathbf{P}(M_y, M_u)$, using some linear fractional algebra manipulations.

The residual generation problem can now be formulated in a pure H_∞ framework by combining both requirements (7.25) and (7.24) into a single H_∞ constraint, that is,

$$\left\| F_l(\tilde{P}(M_y, M_u), F)\right\|_\infty < 1. \tag{7.26}$$

The following proposition, which is an adaptation of proposition 4 in [27] to our problem, gives the SDP (semi-definite programming) solution of Eq. (7.26).

Proposition 7.1 *Let the state–space realization of the transfer $P\tilde{}(M_y, M_u)$ be denoted*

$$\left[\begin{array}{c|cc} \tilde{A} & \tilde{B}_1 & \tilde{B}_2 \\ \hline \tilde{C}_1 & \tilde{D}_{11} & \tilde{D}_{12} \\ \tilde{C}_2 & \tilde{D}_{21} & \tilde{D}_{22} \end{array}\right]$$

so that it is partitioned in accordance with the diagram 7.7b. By construction both $\tilde{C}_1$ and $\tilde{D}_{11}$ depend on M_y, M_u. Furthermore, $\tilde{B}_2 = 0$ and $\tilde{D}_{22} = 0$, showing that the fault detector operates in open loop versus the system (no effect neither on its state nor on its output). Let $W = \left(\tilde{C}_2\tilde{D}_{21}\right)^{\perp}$. Then there exists a solution to (7.26), if and only if there exist $\gamma < 1$, a scalar h, matrices $M_y \in R^{1\times 9}$, $M_u \in R^{1\times 12}$, and two positive definite symmetric matrices R, S solving the following SDP problem:

$\min_\gamma$ s.t.

$$\left(\begin{pmatrix} \begin{pmatrix} I & 0 \\ 0 & h \\ 0 & -h \end{pmatrix} & 0 \\ 0 & I \end{pmatrix}\right)^{\mathrm{T}} \begin{pmatrix} \tilde{A}R + R\tilde{A}^{\mathrm{T}} & R\tilde{C}_1^{\mathrm{T}}(M_y, M_u) & \tilde{B}_1 \\ \tilde{C}_1(M_y, M_u)R & -\gamma I & \tilde{D}_{11}(M_y, M_u) \\ \tilde{B}_1^{\mathrm{T}} & \tilde{D}_{11}^{\mathrm{T}}(M_y, M_u) & -\gamma I \end{pmatrix} \left(\begin{pmatrix} \begin{pmatrix} I & 0 \\ 0 & h \\ 0 & -h \end{pmatrix} & 0 \\ 0 & I \end{pmatrix}\right) < 0 \tag{7.27}$$

$$\begin{pmatrix} W & 0 \\ 0 & I \end{pmatrix}^{\mathrm{T}} \begin{pmatrix} \tilde{A}^{\mathrm{T}}S + S\tilde{A} & S\tilde{B}_1 & \tilde{C}_1^{\mathrm{T}}(M_y, M_u) \\ \tilde{B}_1^{\mathrm{T}}S & -\gamma I & \tilde{D}_{11}^{\mathrm{T}}(M_y, M_u) \\ \tilde{C}_1(M_y, M_u) & \tilde{D}_{11}(M_y, M_u) & -\gamma I \end{pmatrix} \begin{pmatrix} W & 0 \\ 0 & I \end{pmatrix} < 0 \tag{7.28}$$

$$\begin{pmatrix} R & I \\ I & S \end{pmatrix} \geq 0. \tag{7.29}$$

Since γ enters linearly in (7.27) and (7.28), it can be directly minimized by SDP optimization techniques. This enables to find the smallest achievable H_∞-norm and to determine the optimal solution (M_y, M_u, F (s)); see [27, 46] for a procedure to compute the state–space matrices AF, BF, CF, DF of $F\ (s) = CF\ (sI - AF)_{-1}\ BF + DF$ based on linear algebra. However, it is better to use an SDP-based procedure with adequate change of LMI variables if it is required to add additional constraints on the filter such as robust poles assignment (see [28] and Sect. 7.6 for an illustration of the technique).

7.5.3 Computational Results

The SDPT3 solver is used to perform the optimization problem (7.27) (7.28), and (7.29). For each synthesis, the numerical values of γ_{acc}, γ_{att}, and γ_h have been

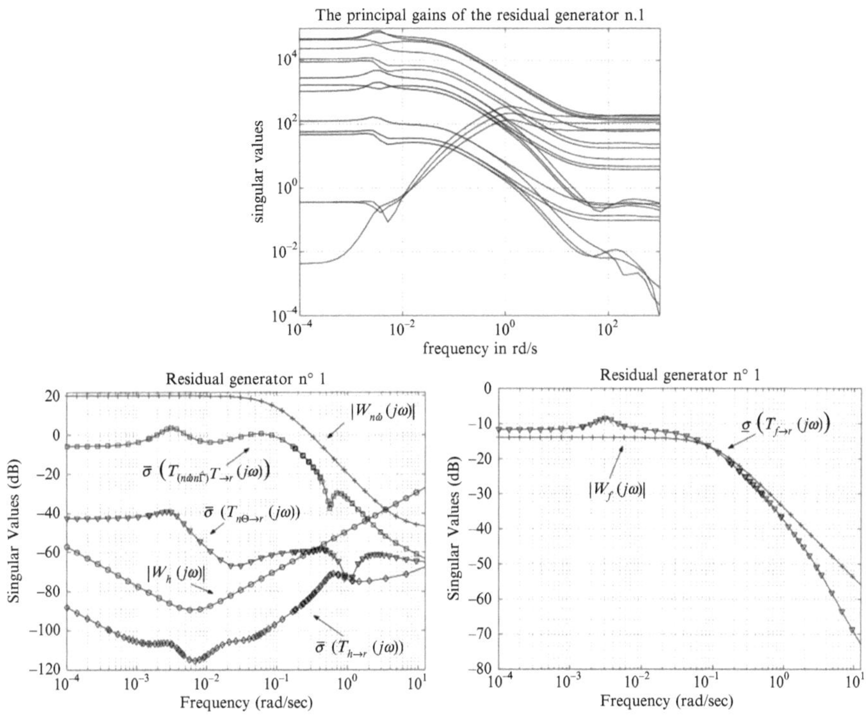

Fig. 7.8 The principal gains of the first residual generator (*top*) and the principal gain of $T^k_{d\to r}$ and $T_{f\to r}$ versus W^k_d and W_f (*bottom*)

fixed to 10^{-2}, 0.1, and 3.10^{-5}, respectively. Note that these small values indicate a high-robustness level of the residual generators against n and $h(\varpi_\alpha, \varpi_{\text{spin}})$ since they indicate the attenuation level of $h(\varpi_\alpha, \varpi_{\text{spin}})$ and n on $r_i\ (t)$, $i = 1, \ldots, 12$. Furthermore, the numerical values of γ_2 and ω_f are maximized in each case. This permits to achieve high sensitivity performances of the detection filters. Figure 7.8 illustrates the principal gains of the first residual generator. For brevity, the other fault detection filters are not considered.

To analyze the computed residual generators, the principal gains $\bar{\sigma}\left(T^k_{d\to r}(j\omega)\right)$ and $\underline{\sigma}\left(T_{f\to r}(j\omega)\right)$ are plotted versus the objectives W^k_d and W_f. The notation "k" is introduced to outline that the analysis is performed with respect to $h(\varpi_\alpha, \varpi_{\text{spin}})$ and each component of n. Figure 7.8 illustrates the plots for the first residual. As it can be seen from the figure, $\bar{\sigma}\left(T^k_{d\to r}(j\omega)\right) < |W^k_d(j\omega)|, \forall\omega$ and $\underline{\sigma}\left(T_{f\to r}(j\omega)\right) > |W_f(j\omega)|, \forall\omega \in \Omega \approx [0; 0.1]\text{rad/s}$ which indicate that the requirements (S.1) and (S.2) are satisfied. Furthermore, the small gap between $\bar{\sigma}\left(T^k_{d\to r}(j\omega)\right)$ and $|W^k_d(j\omega)|, \forall\omega$ and between $\underline{\sigma}\left(T_{f\to r}(j\omega)\right)$ and $|W_f(j\omega)|, \forall\omega \in \Omega$ illustrates a not too conservative solution.

7.5.4 Isolation Strategy

After fault detection and confirmation, fault isolation is required to get a deeper insight into the faulty situation and to identify which FEEP thruster is faulty. The proposed isolation strategy is based on the following cross-correlation criterion between the residuals r_i and the associated controlled thrusters open rate T_i

$$\varrho_i(\tau) = \left| \frac{1}{N} \sum_{k=\tau}^{\tau+N} (r_i(k) - \bar{r}_i) \left(T_i(k) - \bar{T}_i\right) \right|, \quad i = 1, \ldots, 12. \tag{7.30}$$

$\bar{r}_i$ and $\bar{T}_i$ denote the mean value of $r_i(k), k = \tau, \ldots, \tau + N$ and $T_i(k), k = \tau, \ldots, \tau + N$, respectively. The isolation procedure works in such a way that when $\varrho_K(\tau)$ is higher or equal to a prescribed value ρ, then the fault is declared to be localized in the kth actuator. For real-time implementation, this criterion is computed on an N-length sliding window. It should be pointed out that such an isolation strategy makes sense since the sensitivity of the ith residual with respect to the ith FEEP thruster fault has been maximized, in the H_{-}-norm sense.

7.5.5 Nonlinear Simulation Results

The 12 detection filters are converted into discrete time using a Tustin approximation and implemented within the nonlinear simulator of MICROSCOPE. The simulations have been performed on an orbital period, i.e., $t = 0 \ldots 6{,}000$ s. To make a final decision about the fault, a sequential Wald decision test applied to $||r(t)||_2$ is implemented within the simulator. The probabilities of non-detection and false alarms have been fixed to 0.1 %. The isolation strategy has been also implemented. Figure 7.9 illustrates the behavior of the residuals r_i (t), $i = 1, \ldots,$ 12, the behavior of the decision test, and the isolation criteria $\varrho(\tau)$ for some faulty situations. As it can be seen from the figures, after a small transient behavior, all faults are successfully detected and isolated by the FDI unit. Finally, the case of multiple faults is considered. The simulated scenario corresponds to the first FEEP thruster simultaneously closing itself and the fourth blocking itself. Figure 7.10 illustrates the behavior of the residuals $r1$ (t) and $r4$ (t) and the decision test. The behavior of the isolation criteria $\varrho(\tau)$ is also illustrated. As it can be seen, the two faults are successfully detected and isolated.

7.6 A Deep Space Mission: Mars Sample Return

The Mars Sample Return (MSR) mission is a space mission undertaken jointly by NASA and ESA. It is due to be launched by 2020. The goal is to return tangible samples from Mars atmosphere and ground to Earth for analysis. Five spacecraft

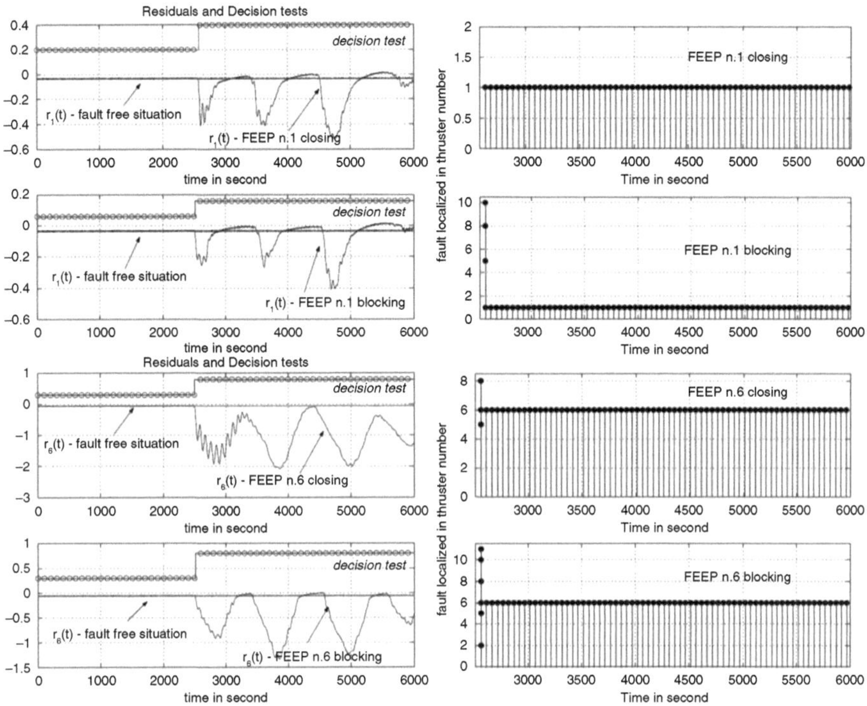

Fig. 7.9 Behavior of the residuals in fault-free and faulty situations + the decision test (*left*) and the isolation criteria (*right*)

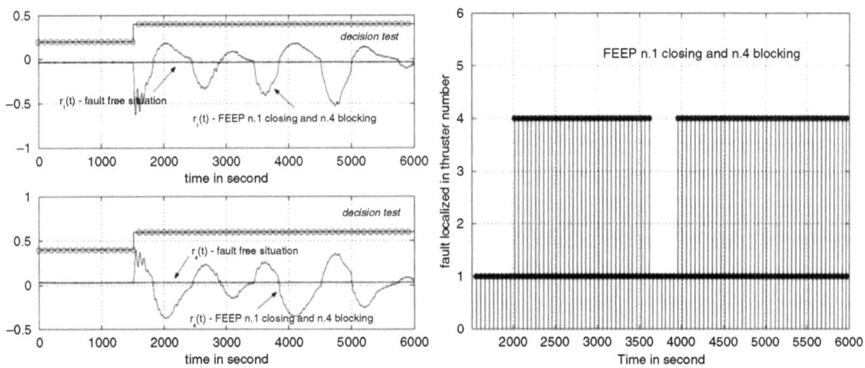

Fig. 7.10 Behavior of the residuals in fault-free and faulty situations + the decision test (*left*) and the isolation criteria (*right*). Thruster n.1 closing and n.4 blocking

are involved within this spatial mission: an Earth/Mars transfer vehicle, an orbiter, a Mars descent vehicle (MDV), a Mars ascent vehicle (MAV), and an Earth reentry vehicle (Fig. 7.11).

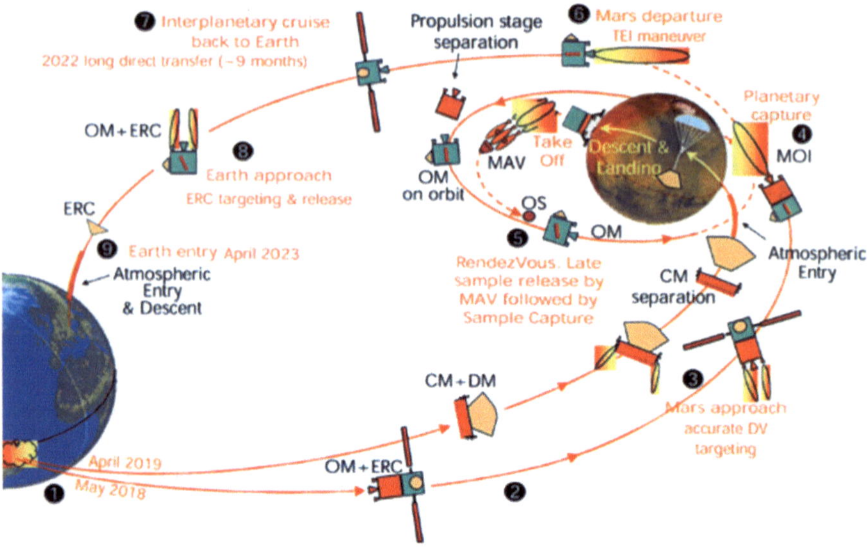

Fig. 7.11 The MSR mission

When the orbiter is in a low-altitude position around the Martian orbit, the MDV is released on the Martian ground. Once the sample collecting process is completed, the samples are loaded on the MAV which is then launched into the Martian orbit around the planet to rendezvous with the orbiter (see Fig. 7.11 for an illustration of the MSR mission). The work reported in this section focuses on the rendezvous phase.

To carry out the rendezvous mission, the orbiter vehicle uses:

(a) A large range of sensors, namely, inertial measurement units (IMUs) containing two 3-axis gyroscopes (GYR) and two 3-axis accelerometers (ACC), two star trackers (STR), two coarse sun sensors (CSS), two global navigation satellite system (GNSS) sensors, two radio frequency sensors (RFS), a light detection and ranging (LIDAR) sensor, and a narrow-angle camera (NAC).
(b) An actuation system composed of a two sets of eight thrusters (THR) and a set of six reaction wheels (RW). This actuation system is designed so that it is possible either to control the position using one set of thruster (the second one being available for FDIR purpose) and the attitude using the reaction wheels or to switch off the RW actuation so that both the position and attitude are controlled by the 2×8 thrusters, thanks to an advanced allocation unit.

The subsequent developments consider the case of attitude control by means of the RW actuation system. However, it should be noted that the presented results are valid in case of controlling the attitude with the thrusters.

The aim of IMU unit is to measure the angular velocities $(p,\ q,\ r)$ and accelerations $(\dot{p}, \dot{q}, \dot{r})$. The STR device gives the attitude measurement $\left(\theta_x, \theta_y, \theta_z\right)$.

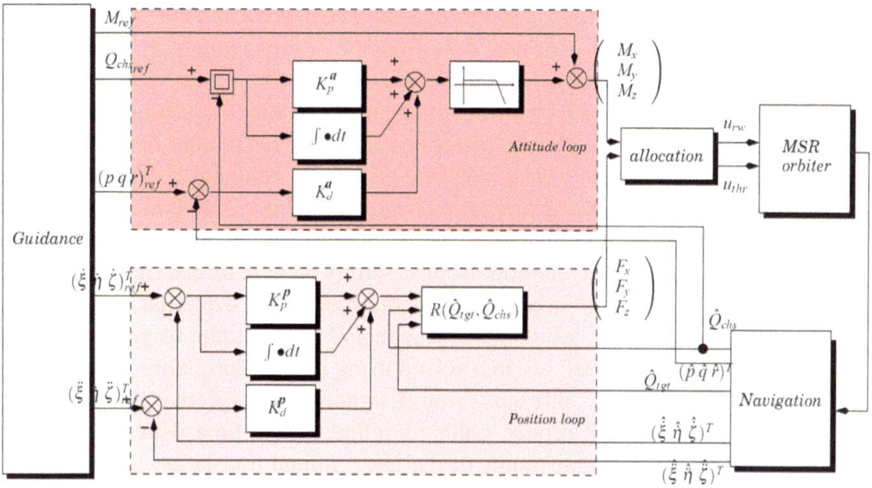

Fig. 7.12 The structure of the orbiter's GNC

The RFS, NAC, and LIDAR units measure the relative position (ξ, η, ζ) between the Mars ascent vehicle and the orbiter at different steps of the rendezvous phase. For instance, the LIDAR is used during the final rendezvous phase, whereas the NAC is used by the orbiter to look for the position of the Mars ascent vehicle around the planet. Note that the CSS and GNSS units measure the position of the spacecraft in space and are not used during the rendezvous phase.

Figure 7.12 presents a diagram of the orbiter GNC (Guidance, Navigation, and Control) system. As classically in aerospace, the GNC consists of (see also Chap. 2):

- A guidance loop that is in charge of computing the rendezvous trajectory, i.e., the desired quaternion of attitude $Q_{\text{chs}_{\text{ref}}}$ of the orbiter, the desired angular velocity $(pqr)^{\text{T}}_{\text{ref}}$, the desired control moments M_{ref}, and the relative velocity $(\dot{\xi}\dot{\eta}\dot{\zeta})^{\text{T}}_{\text{ref}}$ between the target and the ascent vehicle.
- A navigation unit in charge of computing/estimating necessary signals, e.g., the quaternion of the orbiter $\hat{Q}_{\text{chs}}$ and the target $\hat{Q}_{\text{tgt}}$, the pitch, yaw, roll rates $p\ q\ r$, the relative position $\xi\eta\zeta$, velocities $\dot{\xi}\dot{\eta}\dot{\zeta}$, and accelerations $\ddot{\xi}\ddot{\eta}\ddot{\zeta}$.
- A control unit composed of two control laws: (1) the orbit control loop which is composed of PID controller, a rotation matrix $R\left(\hat{Q}_{\text{tgt}}(t), \hat{Q}_{\text{chs}}(t)\right)$, and a THR management unit and (2) an attitude control loop which is composed of a PID controller, high-frequency filters, and an RW management function. The THR and RW management functions play the role of allocation schemes. The first control loop aims at regulating the position of the orbiter through the "open–close" movement of the thrusters (the input signal is denoted u_{thr} on Fig. 7.12), while the second loop provides a controlled attitude of the orbiter through the reaction wheels (the input signal is denoted u_{rw} on Fig. 7.12).

Obviously, the rendezvous mission can be in danger if a fault occurs in the thrusters since the GNC system may not compensate, for example, J_2 disturbances, and/or may lose the attitude and/or the position of the ascent vehicle even if the design of the GNC unit is made more or less robust to some disturbances and faults. Such faulty situations cannot be diagnosed by ground operators using telemetry information due to the potential lack of communication between the orbiter and the ground stations or due to significant delays. This problem becomes especially critical during the last 20 m of the rendezvous phase, since one has to correctly position the orbiter in order to successfully capture the ascent vehicle. The objective here is to present an advanced model-based fault detection and isolation scheme based on robust poles assignment with a $H(0)$ constraint to guarantee fault sensitivity. The isolation task is again solved using a cross-correlation test between the residual signal and the thrusters control signals. The objective is to diagnose thruster faults of the MSR orbiter, onboard/online, and in time within the critical dynamic and operational constraints of the last terminal translation (last 20 m) of the MSR rendezvous/capture phase. During this scenario, the chaser stays in the rendezvous/capture corridor, such that it would be possible to anticipate necessary recovery actions to successfully meet the capture phase (see Fig. 7.11).

Different fault profiles are considered: locked-opened/closed thruster failure, cyclic forces/torques around the desired force/torque profile with small magnitude and monopropellant leakage. However, the subsequent developments will present only the results for locked-opened thruster failures. For instance, a thruster locked closed is more difficult to diagnose because the thruster is not necessarily used at the time of the failure, and because the thrusters, when they are used, produce small pulses whose effect averaged over the control cycle is small. Such faults are highly non-detectable using the standard industrial onboard FDIR techniques and/or ground analysis. Moreover, the uncertainty on the center of mass due to propellant motions in the tanks makes the detection and isolation more challenging.

The key feature of the proposed method is the use of a judiciously chosen linear model for the design of the fault detector, i.e., the model consists of a position model given in a judiciously chosen frame that takes into account both the rotational and linear translation spacecraft motions. In other words, the dynamics of the attitude of the orbiter is not modeled. The model used for designing the fault detector relies only on the dynamics of the relative position between the orbiter and the MAV.

7.6.1 Modeling the Orbiter Dynamics During the Rendezvous Phase

The translation motion of the orbiter is derived from the second Newton law. To proceed, let a, m, G, and m_M denote the orbit of the target, the mass of the orbiter, the gravitational constant, and the mass of the planet Mars. Then, the orbit of the rendezvous being circular, the velocity of any object (e.g., the chaser and the target),

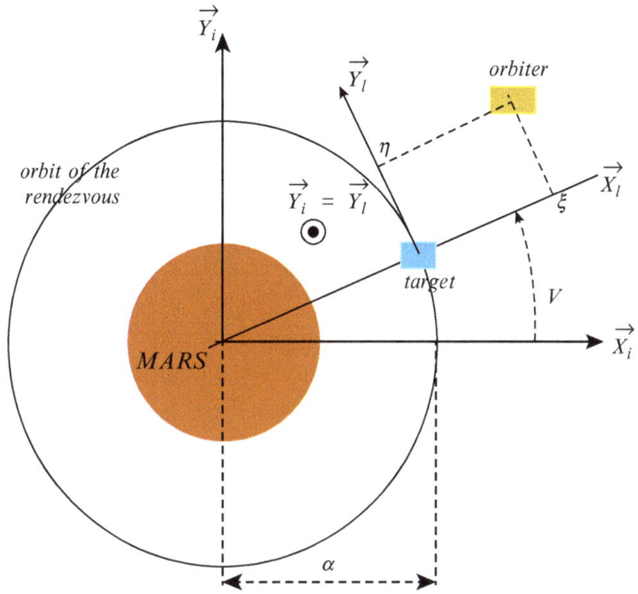

Fig. 7.13 Rendezvous orbit and associated frames

is given by the relation $\sqrt{\frac{\mu}{a}}$ where $\mu = \text{G}\,.mM$ (see, for instance, [37, 38, 47–49]). Let $R_l : (O_{\text{tgt}}, \overrightarrow{X_l}, \overrightarrow{Y_l}, \overrightarrow{Z_l})$ be the frame attached to the target and oriented as shown in Fig. 7.13. Because the linear velocity of the target is given by the relation $a\dot{\theta}$ in the inertial frame $R_i : (O_M, \overrightarrow{X_i}, \overrightarrow{Y_i}, \overrightarrow{Y_i})$ (those that are attached to the center of Mars, see Fig. 7.13), it follows

$$a.\dot{\theta} = \sqrt{\frac{\mu}{a}} \Rightarrow n = \sqrt{\frac{\mu}{a^3}}. \tag{7.31}$$

During the rendezvous phase, it is assumed that the orbiter motion is due to the four following forces:

- The Mars attraction force $\overrightarrow{F_a}$ given in R_l by $\overrightarrow{F_a} = -m\frac{\mu}{\left((a+\xi)^2+\eta^2+\zeta^2\right)^{3/2}}\left((a+\xi)\overrightarrow{X_l}\right.$ $\left.+\eta\overrightarrow{Y_l} + \zeta\overrightarrow{Z_l}\right)$ where ξ, η, ζ denote the three-dimensional position of the orbiter (assumed to be a punctual mass) in R_l
- The inertial force $\overrightarrow{F_e} = m\left(n^2\xi\overrightarrow{X_l} + n^2\eta\overrightarrow{Y_l} + 0\overrightarrow{Z_l}\right)$
- The Coriolis force $\overrightarrow{F_c}$ that is given in R_l by $\overrightarrow{F_c} = m\left(2n\dot{\eta}\overrightarrow{X_l} - 2n\dot{\xi}\overrightarrow{Y_l} + 0\overrightarrow{Z_l}\right)$
- The forces due to the thrusters $\overrightarrow{F_{\text{th}}} = F_\xi\overrightarrow{X_l} + F_\eta\overrightarrow{Y_l} + F_\zeta\overrightarrow{Z_l}$

Then, from the second Newton law, it follows

$$\begin{aligned}\ddot{\xi} &= n^2\xi + 2n\dot{\eta} - \frac{\mu}{\left((a+\xi)^2+\eta^2+\zeta^2\right)^{3/2}}(a+\xi) + \frac{F_\xi}{m}\\ \ddot{\eta} &= n^2\eta - 2n\dot{\xi} - \frac{\mu}{\left((a+\xi)^2+\eta^2+\zeta^2\right)^{3/2}}\eta + \frac{F_\eta}{m}\\ \ddot{\zeta} &= -\frac{\mu}{\left((a+\xi)^2+\eta^2+\zeta^2\right)^{3/2}}\zeta + \frac{F_\zeta}{m}.\end{aligned} \tag{7.32}$$

Because the distance between the target and the orbiter is smaller than the orbit a, it is possible to derive the so-called Hill–Clohessy–Wiltshire equations [37, 38, 47–49] from Eq. (7.32) by means of a first-order approximation. This gives a six-order linear state–space model whose input vector is $F = (F_\xi F_\eta F_\zeta)^{\mathrm{T}}$ and state vector is $x = (\xi\eta\zeta\dot{\xi}\dot{\eta}\dot{\zeta})^{\mathrm{T}}$. Now, projecting the thrust forces due to the eight thrusters that equip the orbiter into the frame R_l, it follows from (7.32)

$$\begin{cases} \dot{x} = Ax + BR(\hat{Q}_{\mathrm{tgt}}(t), \hat{Q}_{\mathrm{chs}}(t))M u_{\mathrm{thr}}(t) + B_w w(t) \\ y = Cx + n. \end{cases} \tag{7.33}$$

In (7.33), $\hat{Q}_{\mathrm{tgt}} \in R^4$ and $\hat{Q}_{\mathrm{chs}} \in R^4$ denote the attitude's quaternion of the target and the orbiter, respectively. These quaternions are denoted as estimates since they are provided (i.e., estimated) by the navigation module. $M\in\mathrm{R}^{3\times 8}$ refers to the thrust direction matrix. $u_{thr}\in R^8$ refers to the thrusters input and $R(.)$ is used for a rotation matrix. $x\in R^6$ is the state vector defined previously, $y\in R^3$ refers to the three-dimensional positions measured by the LIDAR unit, and $w\in R^3$ refers to the spatial disturbances. The considered disturbances in this study are solar radiations, gravity gradient, and atmospheric drag. n denotes the measurement noise assumed to be a white noise with very small variance due to the technology used for the design of the LIDAR. A, B, and C are matrices of adequate dimension.

Thruster faults can be modeled in a multiplicative manner according to (the index f is again used to outline the faulty case)

$$u_{\mathrm{thr}}^f(t) = (I_8 - \Psi(t))u_{\mathrm{thr}}(t), \quad \Psi(t) = \mathrm{diag}\{\psi_i(t)\} : 0 \le \psi_i(t) \le 1, i = 1, \ldots, 8 \tag{7.34}$$

where $\psi_i, i = 1, \ldots, 8$ are unknown. $\Psi(t)$ models the thruster faults, e.g., a locked-in-place fault in the ith thruster can be modeled by $\psi_i(t) = 1 - \frac{c}{u_{\mathrm{thr}_i}(t)}$ where c denotes a constant value (the particular values $c = \{0, 1\}$ are used to consider open/closed faults), whereas a fix value of ψ_i models a loss of efficiency of the ith thruster. $\Psi(t) = 0, \forall t$ means that no fault occurs in the thrusters.

Then taking into account unknown but bounded delays induced by the electronic devices and the uncertainties on the thruster rise times due to the thruster modulator unit that is modeled here to be a constant gain with unknown but bounded time delay $\tau = \tau_0 \pm \delta_\tau : |\delta_\tau| \leq \overline{\delta_\tau}$, the motion of the orbiter during the rendezvous can be modeled in both fault-free (i.e., $\Psi = 0$) and faulty (i.e., $\Psi \neq 0$) situations according to

$$\begin{cases} \dot{x} = Ax + BR(\hat{Q}_{\text{tgt}}(t), \hat{Q}_{\text{chs}}(t))M(I - \Psi(t))u_{\text{thr}}(t - \tau) + B_w w(t) \\ y = Cx + n \end{cases}$$

$$\tau = \tau_0 \pm \delta_\tau : |\delta_\tau| \leq \overline{\delta_\tau}\Psi(t) = \text{diag}(\psi_i(t)) : 0 \leq \psi_i(t) \leq 1, i = 1, \ldots, 8. \tag{7.35}$$

Using a Padé approximation of the time delay τ and approximating the fault model $R(\hat{Q}_{\text{tgt}}(t), \hat{Q}_{\text{chs}}(t))M\Psi(t)u_{\text{thr}}(t)$ in terms of additive faults $f(t) \in R_3$ acting on the state via a constant distribution matrix K_f (then $K_f = B$) and defining the generalized input vector u as

$$u(t) = R(\hat{Q}_{\text{tgt}}(t), \hat{Q}_{\text{chs}}(t))M\, u_{\text{thr}}(t). \tag{7.36}$$

It follows that the overall model of the orbiter dynamics that takes into account both the attitude ($Q_{\text{chs}}(t)$) and the relative position ($x(t)$) of the orbiter can be written in a LFR form, i.e., the uncertain parameter τ is "pulled out" so that (7.35) appears as a nominal model P subject to an artificial feedback $\Delta = \delta_\tau I_8$, that is,

$$y(s) = F_u\left(P(s), \Delta\right) \begin{pmatrix} w(s) \\ f(s) \\ u(s) \end{pmatrix} + n(s), \quad \Delta = \delta_\tau I_8 : ||\Delta|| \leq 1. \tag{7.37}$$

7.6.2 Design of the FDI System

The FDI design follows the general theory presented in [28]. The general setup of the FDI scheme is illustrated on Fig. 7.14. It consists of a residual generator r so that

$$r(s) = F(s) \begin{pmatrix} y(s) \\ u(s) \end{pmatrix} \quad r \in R^3 \tag{7.38}$$

which is designed so that the sensitivity level of the residual with respect to any thruster fault f is maximized in the $H(0)$-norm sense while guaranteeing robustness against the noise n, the spatial disturbances w for the considered uncertainty block Δ. Such robustness requirements are managed through robust poles assignment.

To proceed, let the LFR model $F_u(P(s), \Delta)$ be robustly stable (this can be done without loss of generality since P may include the position controller K_{pos}; see

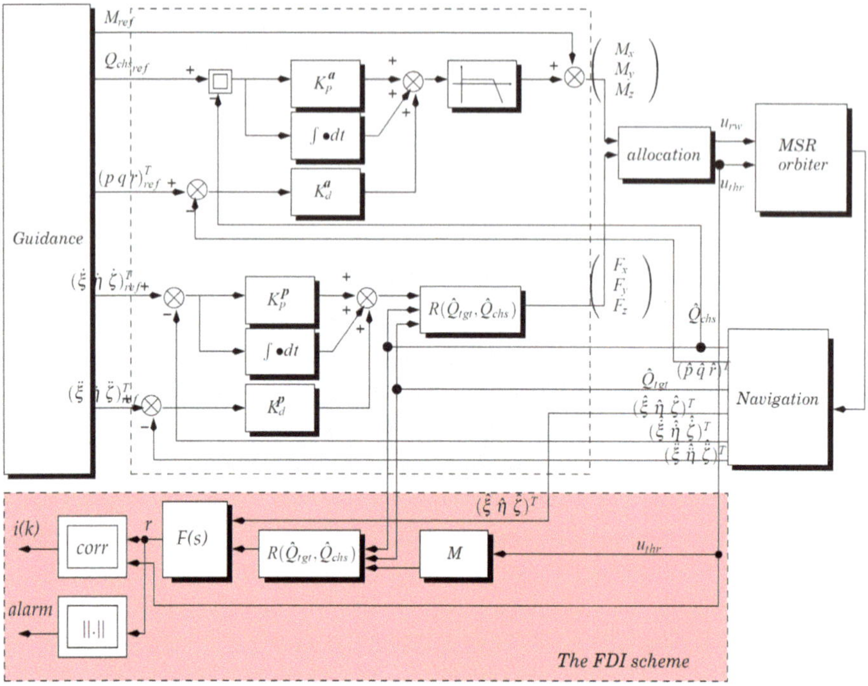

Fig. 7.14 The FDI scheme for thruster fault diagnosis in the orbiter spacecraft

Fig. 7.12) and the fault f_i be observable from the output y. These assumptions are prior conditions for the fault detection problem being well posed. Consider the residual vector r defined by Eq. (7.38). The residual generator design problem can be formulated as the design of a filter $F(s) = C_F\,(sI - A_F)^{-1}B_F + D_F$ that solves the following optimization problem:

$$\begin{array}{lll} \max_{A_F,B_F,C_F,D_F} & \varphi & \\ s.t. & ||T_{f\rightarrow r}||_0 > \varphi & ,\forall\Delta : ||\Delta||_\infty \leq 1 \\ & \lambda_i(A_F) \in \mathcal{R} \subseteq D, \forall i. & \end{array} \tag{7.39}$$

$T_{f\rightarrow r}$ denotes the transfer between f and r and D denotes the left half-complex plane. λ_i refers to the ith eigenvalue of the matrix A_F, and φ denotes the fault sensitivity performance index for the residual vector (7.38).

The constraint $\lambda_i(A_F) \in \mathcal{R} \subseteq D$, $\forall i$ refers to a robust pole assignment constraint and the performance index φ guarantees a maximum faults amplification $H(0)$ gain. In other words, the problem is formulated so that the robustness requirements against Δ, w, and n are specified through $\mathcal{R}$ while specifying a high fault sensitivity level of the residual vector r through the maximization of φ. Thus,

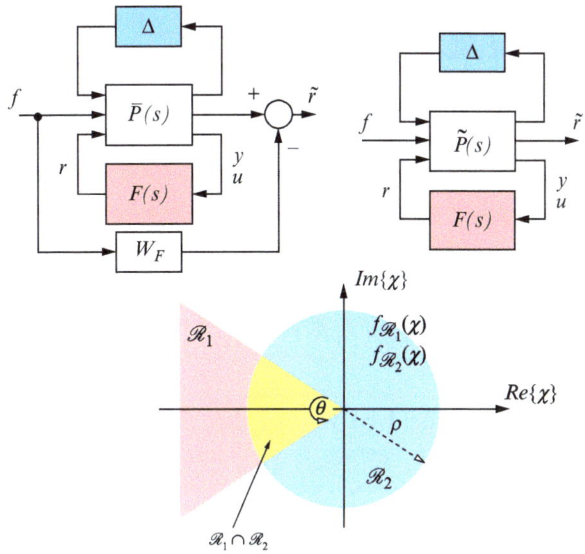

Fig. 7.15 Fault detector design problem: $H(0)$ (*left*) and poles assignment (*right*) specifications

n and w are ignored from now on. Note that, in practice, $\mathcal{R}$ is a parameter to be selected by the designer, since finding an optimal region for $\mathcal{R}$ that guarantees high nuisances rejection is highly related to the system under consideration.

7.6.2.1 The SDP Formulation of the H (0) Specification

Similarly to the case of the MICROSCOPE experiment (see Sect. 7.5), a shaping filter W_f is used to specify the fault sensitivity objectives. The solution of the $H(0)$ specification (7.39) is then handled using the following lemma, which is a direct application of lemma 1 taking into account the definition of the $H(0)$ gain.

Lemma 7.2 *Let W_f be defined so that $||W_f||_0 \neq 0$. Introduce W_F, a right invertible transfer matrix, so that $||W_f||_0 = \frac{\varphi}{\alpha}||W_F||_0$ and $||W_F||_0 > \alpha$ where $\alpha = 1 + \varphi$. Define the signal $\tilde{r}$ such that $\tilde{r}(s) = r(s) - W_F\,(s)\,f\,(s)$: $\tilde{r}$ (see Fig. 7.15 for easy reference). Then a sufficient condition for the H (0) specification in (7.39) to hold is*

$$||T_{f\to\tilde{r}}||_\infty < 1, \quad \forall\Delta : ||\Delta||_\infty \leq 1 \tag{7.40}$$

where $T_{f\to\tilde{r}}$ denotes the closed-loop transfer between $\tilde{r}$ and f.

Using Lemma 7.2, the $H(0)$ specification can be transformed to a fictitious H_∞ framework: With W_F and using some LFR algebra, one can derive from (7.37) and (7.38) a new model $P\tilde{}$ such that (see Fig. 7.15 for easy reference)

$$\tilde{r}(s) = F_u\left(F_l\left(\tilde{P}(s), F(s)\right), \Delta\right) f(s). \tag{7.41}$$

Noting that $F_u\left(F_l\left(\tilde{P}(s), F(s)\right), \Delta\right)$ is the transfer $T_{f\to\tilde{r}}$, it follows, by virtue of the small gain theorem [50], that a sufficient condition for the $H(0)$ specification to hold is

$$\exists F(s) : \left\| F_l\left(\tilde{P}(s), F(s)\right)\right\|_{\infty} < 1. \tag{7.42}$$

Let $\left(\tilde{A}, \tilde{B}, \tilde{C}, \tilde{D}\right)$ be the state–space matrices of $\tilde{P}$ and consider the following partition of $\tilde{B}, \tilde{C}$, and $\tilde{D}$accordance with the diagram depicted on Fig. 7.15:

$$\tilde{B} = \left(\tilde{B}_1 \tilde{B}_2\right), \quad \tilde{C} = \begin{pmatrix} \tilde{C}_1 \\ \tilde{C}_2 \end{pmatrix}, \quad \tilde{D} = \begin{pmatrix} \tilde{D}_{11} & \tilde{D}_{12} \\ \tilde{D}_{21} & \tilde{D}_{22} \end{pmatrix}. \tag{7.43}$$

Again, it could be verified that $\tilde{B}_2 = 0$ and $\tilde{D}_{22} = 0$, showing that the fault detection filter F operates in open loop versus the system. Then, using some linear algebra manipulations, it can be verified that the closed-loop model $F_l\left(\tilde{P}(s), F(s)\right)$ admits the state realization (A_c, B_c, C_c, D_c) which is deduced from those of $P^{\sim}$ and F:

$$A_c = \begin{pmatrix} \tilde{A} & 0 \\ B_F\tilde{C}_2 & A_F \end{pmatrix}, B_c = \begin{pmatrix} \tilde{B}_1 \\ B_F\tilde{D}_{21} \end{pmatrix}, C_c = \left(\tilde{C}_1 + \tilde{D}_{12}D_F\tilde{C}_2 \;\; \tilde{D}_{12}C_F\right)$$
$$D_c = \tilde{D}_{11} + \tilde{D}_{12}D_F\tilde{D}_{21}. \tag{7.44}$$

From [51], $F_l\left(\tilde{P}(s), F(s)\right)$ is stable (and F is a robustly stable filter due to the triangular structure of A_c) and there exists a solution to (7.42) if and only if there exists $\gamma < 1$ and matrices $\mathbf{A}$, $\mathbf{B}$, $\mathbf{C}$, $\mathbf{D}$, $\mathbf{X} = \mathbf{X}^{\mathrm{T}}$ and $\mathbf{Y} = \mathbf{Y}^{\mathrm{T}}$ that solve the following SDP problem:

$\min_{\gamma}$ s.t.

$$\begin{pmatrix} \tilde{A}\mathbf{X} + \mathbf{X}\tilde{A}^{\mathrm{T}} & \mathbf{A}^{\mathrm{T}} + \tilde{A} & \tilde{B}_1 & \left(\tilde{C}_1\mathbf{X} + \tilde{D}_{12}\mathbf{C}\right)^{\mathrm{T}} \\ \mathbf{A} + \tilde{A}^{\mathrm{T}} & \tilde{A}^{\mathrm{T}}\mathbf{Y} + \mathbf{Y}\tilde{A} + \mathbf{B}\tilde{C}_2 + (\mathbf{B}\tilde{C}_2)^{\mathrm{T}} & \mathbf{Y}\tilde{B}_1 + \mathbf{B}\tilde{D}_{21} & \left(\tilde{C}_1 + \tilde{D}_{12}\mathbf{D}\tilde{C}_2\right)^{\mathrm{T}} \\ \tilde{B}_1^{\mathrm{T}} & \left(\mathbf{Y}\tilde{B}_1 + \mathbf{B}\tilde{D}_{21}\right)^{\mathrm{T}} & -\gamma I & \left(\tilde{D}_{11} + \tilde{D}_{12}\mathbf{D}\tilde{D}_{21}\right)^{\mathrm{T}} \\ \tilde{C}_1\mathbf{X} + \tilde{D}_{12}\mathbf{C} & \tilde{C}_1 + \tilde{D}_{12}\mathbf{D}\tilde{C}_2 & \tilde{D}_{11} + \tilde{D}_{12}\mathbf{D}\tilde{D}_{21} & -\gamma I \end{pmatrix} < 0$$
$$\begin{pmatrix} \mathbf{X} & I \\ I & \mathbf{Y} \end{pmatrix} > 0. \tag{7.45}$$

The fault detector state–space matrices A_F, B_F, C_F, and D_F are then deduced from $\mathbf{A}$, $\mathbf{B}$, $\mathbf{C}$, $\mathbf{D}$, $\mathbf{X}$, and $\mathbf{Y}$ so that

$$D_F = \mathbf{D}, \quad C_F = (\mathbf{C} - \mathbf{D}\tilde{C}_2\mathbf{X})M^{-\mathrm{T}}, \quad A_F = N^{-1}(\mathbf{A} - NB_F\tilde{C}_2\mathbf{X} - \mathbf{Y}\tilde{A}\mathbf{X}M^{-\mathrm{T}})$$
$$B_F = N^{-1}\,\mathbf{B} \quad MN^{\mathrm{T}} = I - \mathbf{XY}. \tag{7.46}$$

7.6.2.2 SDP Formulation for the Robust Poles Assignment Specification

Consider now the specification $\lambda_i(A_F) \in \mathcal{R} \subseteq D, \forall i$. Assume that the region $\mathcal{R}$ is formed by the intersection of N elementary LMI regions $\mathcal{R}_i$, i.e., $\mathcal{R} = \mathcal{R}_1 \cap \ldots \cap \mathcal{R}_N$ (see Fig. 7.15 for easy reference). Each LMI region $\mathcal{R}_i$ is characterized as follows:

$$\mathcal{R}_i = \left\{ \chi \in C : L_i + \chi Q_i + \chi^* Q_i^{\mathrm{T}} < 0 \right\} \tag{7.47}$$

where L_i and Q_i are real symmetric matrices. The matrix-valued function $f_{\mathcal{R}_i}(\chi) = L_i + \chi Q_i + \chi^* Q_i^{\mathrm{T}}$ is called the characteristic function of the ith LMI region $\mathcal{R}_i$. Then, it is shown in [52] that a sufficient condition for all eigenvalues of A_c given by (7.44) lying in the region $\mathcal{R}$ for all $\Delta \in \underline{\Delta} : ||\Delta||_\infty \leq 1$ boils down to the existence, for each region $\mathcal{R}_i$, of a matrix Pi and $\beta < 1$ so that

$$\begin{pmatrix} \mathbb{Q}(A_c, P_i) & Q_{1i}^{\mathrm{T}} \otimes (P_i B_c) & Q_{2i}^{\mathrm{T}} \otimes C_c^{\mathrm{T}} \\ Q_{1i} \otimes (B_c^{\mathrm{T}} P_i) & -\beta I & I \otimes D_c^{\mathrm{T}} \\ Q_{2i} \otimes C_c & I \otimes D_c & -\beta I \end{pmatrix} < 0, \quad P_i > 0, \quad i = 1 \ldots N \tag{7.48}$$

$$\mathbb{Q}(A_c, P_i) = L_i \otimes P_i + Q_i \otimes (P_i A_c) + Q_i^{\mathrm{T}} \otimes (A_c^{\mathrm{T}} P_i) \tag{7.49}$$

where "$\otimes$" denotes the Kronecker product of matrices. $Q^{\mathrm{T}} Q_{2i} = Q_i$ is a factorization of Q_i so that Q_{1i} and Q_{2i} have full column rank.

Due to the triangular structure of A_c, it is obvious that the set of the eigenvalues of A_c are equal to the set of the eigenvalues of $\tilde{A}$ and A_F. Thus, a sufficient condition for all fault detection filter poles being located in the region $\mathcal{R}$ for all $\Delta \in \underline{\Delta} : ||\Delta||_\infty \leq 1$ (i.e., for the robust pole assignment specification to hold) is the existence of a solution to the inequalities (7.48). Unfortunately, since each inequality constraint involves products of matrix P_i, $i = 1, \ldots, N$ and the fault filter variables A_F, B_F, C_F, D_F, the resulting optimization problem is nonlinear. To reduce the problem to a linear optimization problem, the linearizing change of variables given by (7.46) can be used.

Let $\tilde{B}_1, \tilde{C}_1, \tilde{D}_{11}, \tilde{D}_{12}, \tilde{D}_{21}$ be partitioned according to the dimension of Δ such that

$$\tilde{B}_1 = \begin{pmatrix} B_\Delta & B_f \end{pmatrix}, \tilde{C}_1 = \begin{pmatrix} C_\Delta \\ C_r \end{pmatrix}, \tilde{D}_{11} = \begin{pmatrix} D_{\Delta\Delta} & D_{\Delta f} \\ D_{r\Delta} & D_{rf} \end{pmatrix}, \tilde{D}_{12} = \begin{pmatrix} D_{1\Delta} \\ D_{1r} \end{pmatrix},$$
$$\tilde{D}_{21} = \begin{pmatrix} D_{2\Delta} & D_{2f} \end{pmatrix}. \tag{7.50}$$

It follows that all eigenvalues of A_F lie in the region $\mathcal{R}$ for all $\Delta \in \underline{\Delta} : ||\Delta||_\infty \leq 1$ if there exist **A**, **B**, **C**, **D**, and $\mathbf{X_i} = \mathbf{X_i^T}, \mathbf{Y_i} = \mathbf{Y_i^T}$ $i = 1 \ldots N$ that solve the following SDP problem:

$$\min \beta s.t.$$
$$\begin{pmatrix} L_i \otimes \Psi(\mathbf{X_i}, \mathbf{Y_i}) + Q_i \otimes \Phi_A + Q_i^{\mathrm{T}} \otimes \Phi_A^{\mathrm{T}} & Q_{1i}^{\mathrm{T}} \otimes \Phi_B & Q_{2i}^{\mathrm{T}} \otimes \Phi_C^{\mathrm{T}} \\ Q_{1i} \otimes \Phi_B^{\mathrm{T}} & -\beta I & I \otimes \Phi_D^{\mathrm{T}} \\ Q_{2i} \otimes \Phi_C & I \otimes \Phi_D & -\beta I \end{pmatrix} < 0$$

$$\text{with } \Psi(\mathbf{X_i}, \mathbf{Y_i}) = \begin{pmatrix} \mathbf{X_i} & I \\ I & \mathbf{Y_i} \end{pmatrix} > 0, \Phi_A = \begin{pmatrix} \tilde{A}\mathbf{X}_i & \tilde{A} \\ \mathbf{A} & \mathbf{Y}_i \tilde{A} + \mathbf{B}\tilde{C}_2 \end{pmatrix}$$

$$\Phi_B = \begin{pmatrix} B_\Delta \\ \mathbf{Y}_i B_\Delta + \mathbf{B} D_{2\Delta} \end{pmatrix}$$

$$\Phi_C = \left(C_\Delta \mathbf{X}_i + D_{1\Delta}\mathbf{C} \; C_\Delta + D_{1\Delta}\mathbf{D}\tilde{C}_2 \right), \Phi_D = D_{\Delta\Delta} + D_{1\Delta}\mathbf{D} D_{2\Delta} \tag{7.51}$$

for all $i = 1 \ldots N$.

7.6.3 Computational Issues

From the above developments, once the $H(0)$ fault sensitivity specification and the poles assignment constraints are specified, $F(s)$ is computed by jointly solving the SDP problems (7.45) and (7.51). Here, W_f is chosen to be a low-pass first-order filter with the highest possible static gain (i.e., with $H(0)$ gain). Robust pole clustering in the LMI region is required as the intersection of the two following regions:

- $\mathcal{R}_1$: disk with center $(-0.5, 0)$ and radius 1 (to prevent fast dynamics). This region is defined according to

$$\mathcal{R}_1 = \left\{ \chi \in C : \begin{pmatrix} -1 & 0.5 \\ 0.5 & -1 \end{pmatrix} + \chi \begin{pmatrix} 0 & 1 \\ 0 & 0 \end{pmatrix} + \chi^* \begin{pmatrix} 0 & 0 \\ 1 & 0 \end{pmatrix} < 0 \right\}.$$

- By this choice, it is required that all eigenvalues of A_F to be close to -0.5.
- $\mathcal{R}_2$: shifted conic sector with apex at ω and angle θ, that is,

$$\mathcal{R}_2 = \left\{ \chi \in C : \begin{pmatrix} -2\omega\cos(\theta) & 0 \\ 0 & -2\omega\cos(\theta) \end{pmatrix} + \chi \begin{pmatrix} \cos(\theta) & \sin(\theta) \\ -\sin(\theta) & \cos(\theta) \end{pmatrix} + \ldots \right.$$
$$\left. \ldots + \chi^* \begin{pmatrix} \cos(\theta) & -\sin(\theta) \\ \sin(\theta) & \cos(\theta) \end{pmatrix} < 0 \right\}$$

where the numerical values of ω and θ are fixed, respectively, to $\omega = 10$ and $\theta = 5°$. This particular region is chosen to maintain a suitable damping ratio of $r(t)$. Note that as (7.45) enforces filter stability, it is inconsequential that the LMI region $\mathcal{R}$ intersects the right half-plane.

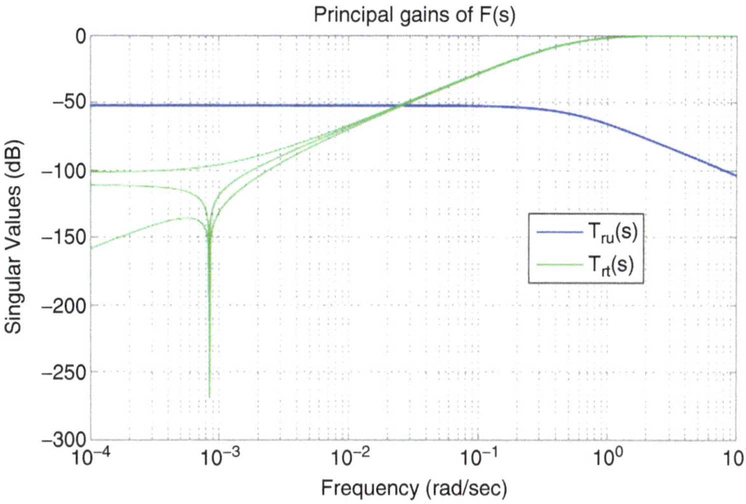

Fig. 7.16 The principal gains of the filter F

We get finally a multi-objective optimization problem in the form

$$\min \varepsilon\gamma + (1+\varepsilon)\beta \quad s.t\ (7.45) \text{ and } (7.51). \tag{7.52}$$

Here, the choice of β is guided by the Pareto optimal points. However, as already mentioned, β is better considered to be a parameter fixed to $\beta = 1$. Thus, the resulting optimization problem looks for the best achievable $H(0)$ objective, whereas the robust pole assignment constraint is enforced. Any $\gamma < 1$ indicates that the obtained solution is admissible. However, $\gamma \approx 1^-$ is required in order to obtain a low conservative solution. Furthermore, all inequalities involved in (7.45) and (7.51) must be solved by using a single Lyapunov matrix for feasibility reasons. This leads to the additional constraints $\mathbf{X}_1 = \mathbf{X}_2 = \mathbf{X}$ and $\mathbf{Y}_1 = \mathbf{Y}_2 = \mathbf{Y}$.

Figure 7.16 illustrates the principal gains $T_{u\to r}\ (j\omega)$ (the transfer between the inputs u and the residuals r) and $T_{y\to r}\ (j\omega)$ (the transfer between the measurements y and the residuals r) of the computed filter F. As it can be seen, $T_{u\to r}\ (j\omega)$ behaves as a low-pass filter, whereas $T_{y\to r}\ (j\omega)$ behaves like a high-pass filter. Furthermore, it can be noted that the gains of $T_{y\to r}\ (j\omega)$ is always lower than 1, showing that the measurement noise is not amplified on the residuals $r(t)$.

7.6.4 Isolation Strategy

For fault isolation, the following normalized cross-correlation criterion between the residuals r and the associated controlled thrusters open rate u_{thr_i} is used here:

$$\sigma(k) = \arg\min_{i} \frac{1}{N} \sum_{k=\tau-N}^{\tau} (r_j(k) - \overline{r_j})(u_{\text{thr}_i}(k) - \overline{u_{\text{thr}_i}}), \quad i = 1\ldots 8, t = k.T_s. \tag{7.53}$$

In (7.53), r_j refers to the jth component of the residual vector r. $\overline{r_j}$, $\overline{u_{\text{thr}_i}}, i = 1\ldots 8$ and T_s denote the mean values of r and u_{thr_i}, $i = 1\ldots 8$ and the navigation module sampling period. For real-time implementation, this criterion is computed on an N-length sliding window. The resulting index $\sigma(k)$ also refers to the identified faulty thruster. A key feature of this isolation strategy is that it is static and then has low computational burdens.

7.6.5 *Nonlinear Simulation Results*

The fault detection filter F is transformed into discrete time using a Tustin approximation and implemented within a highly representative simulator of the MSR mission. The simulated faults correspond to a single thruster opening at 100 % during the last 20 m of the rendezvous. To make a final decision about the fault, a simple threshold is applied on $||r(t)||_2$. The isolation strategy is also implemented with $j = 1$ (see (7.53)). Figure 7.17 illustrates the behavior of the residual $r(t)$ and the isolation criteria $\sigma(t)$ for some faulty situations.

For each simulation, the fault occurs at $t = 100$ s and is maintained. The strategy works as follows: as soon as the fault is declared by the decision test, the cross-correlation criterion (7.53) is computed. As it can be seen from the figures, all thruster faults are successfully detected and isolated by the FDI unit with a detection and isolation delay less than 1.1 s. Note that such a strategy reveals high performance since both the rotational ($Q_{\text{chs}}(t)$) and linear translation ($x(t)$) orbiter motions have been considered as a part of the FDI scheme. In other words, the effects that faults have on both the orbiter attitude and translation motions are taken into account.

7.7 An Atmospheric Reentry Mission

For clarity, in this section some additional specific notations will be used:

α, β = angle-of-attack and sideslip
ϕ, θ = roll and pitch angles
p, q, r = roll, pitch, and yaw rates
x, y, h = ground position of the vehicle
u_b, v_b, w_b = 3-axis velocity components
V_{TAS} = true airspeed

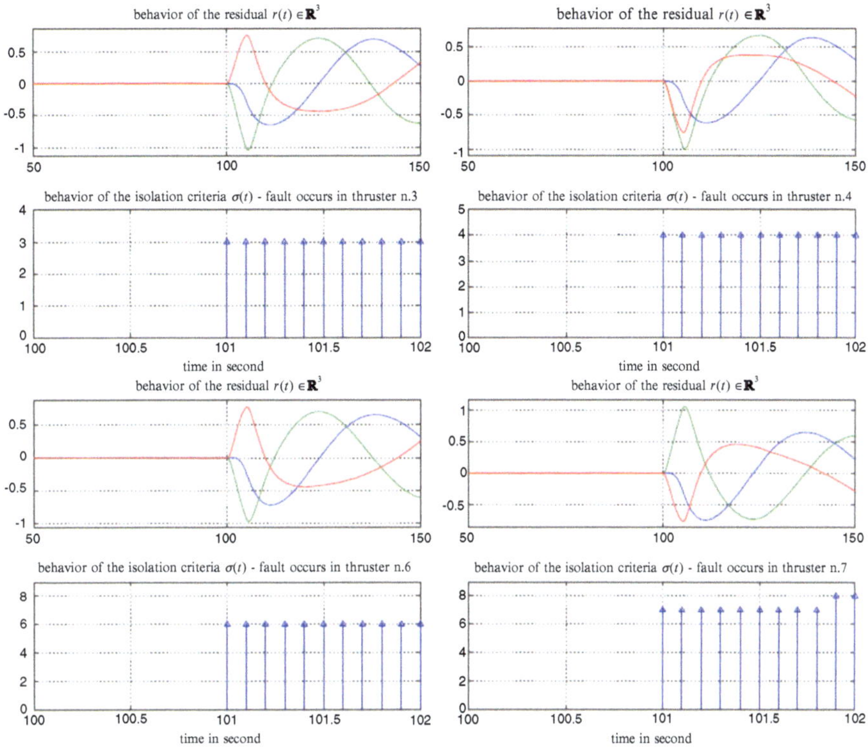

Fig. 7.17 Behavior of $r(t)$ and $\sigma(t)$ for some faulty situations

Mach = Mach number
$\bar{q}$ = dynamic pressure
X_{cg} = center of gravity coordinate
I_{xx}, I_{yy}, I_{zz} = moments of inertia about x-, y-, z-axis
C_{x0} = axial force coefficient
C_{y0} = side force coefficient
C_{z0} = normal force coefficient
C_{l0} = rolling moment coefficient
C_{m0} = pitching moment coefficient
C_{n0} = yawing moment coefficient
δ_{bfll}, δ_{bflr} = lower left and right body-flap deflection
δ_{bful}, δ_{bfur} = upper left and right body-flap deflection
δ_{wfl}, δ_{wfr} = left and right winged-flap deflection
δ_r = rudder deflection
L/D = lift-to-drag ratio
M = vehicle mass

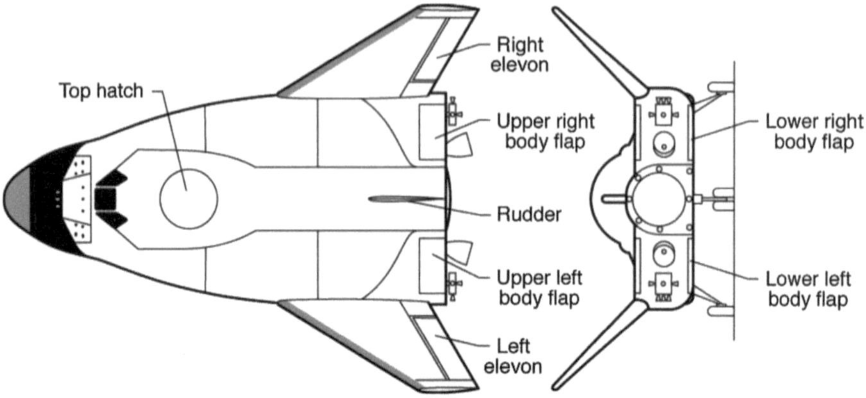

Fig. 7.18 The HL-20 vehicle

A typical atmospheric reentry trajectory for a medium- or high-L/D vehicle has been described in Sect. 2.1.2. See Fig. 2.3. It consists in three successive flight phases, namely, the hypersonic phase from about 120 km-Mach 25 high down to TAEM (Terminal Area Energy Management) handover, the TAEM phase from Mach 2 gate down to Mach 0.5 gate, and the auto-landing phase from Mach 0.5 gate down to the wheel stop on the runway.

The RLV which is considered here is the HL-20 (see Fig. 7.18). This vehicle, defined as a component of the Personnel Launch System (PLS) mission, was initially designed to ensure several manned-space missions including the orbital rescue of astronauts, the International Space Station (ISS) crew exchange, and some satellite repair missions. The dynamics of the HL-20 winged-body vehicle includes a 6 degree-of-freedom mathematical model of the vehicle dynamic, aerologic (wind, atmospheric) models, and a complete GNC (Guidance, Navigation, and Control) architecture dedicated to a specific phase. The guidance algorithm ensures the tracking of the reference trajectory which is computed by the path planner. The guidance commands are the inputs to the control loops which compute the control torques and forces providing conventional rudder (d_r), aileron (d_a), and elevator (d_e) authorities. These signals are next converted into the actuator control input vector $\delta = [\delta wfl\,, \delta wfr\,, \delta bful\,, \delta bfll\,, \delta bfur\,, \delta bflr\,, \delta r\,]^T$ by means of an allocation control algorithm. δ is then applied to the actuators which are modeled as second-order transfer functions. Information on body angular rates is supplied by three rate gyros integrated in an IMU unit. A flush air data system provides information on the angle-of-attack, sideslip, true airspeed velocity, and the dynamic pressure. Information about roll and pitch angles are obtained also through the navigation module. The ground position of the vehicle is finally given by a ground positioning system. It is assumed that all these measurements are corrupted by high-frequency noises (n).

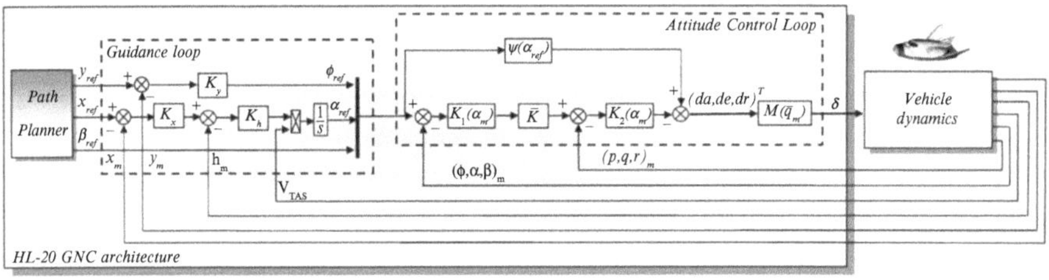

Fig. 7.19 The HL-20 GNC system for the auto-landing phase

7.7.1 The Auto-Landing Phase

7.7.1.1 Modeling the Attitude Control Loop

The general setup of the HL-20 GNC architecture for the auto-landing phase is provided in Fig. 7.19. As classically in the aerospace applications, the flight control system remains in a gain scheduled-based configuration, where the scheduling parameters are the dynamic pressure $\bar{q}_m$ and the angle-of-attack α_m. $K_1(\alpha_m)$ and $K_2(\alpha_m)$ are two varying gains depending on the angle-of-attack $\alpha_{m,}$ and $\bar{K}(s)$ is a dynamical controller designed to ensure stability and keep some performance level. The subscript "m" denotes the fact that information is provided by the navigation module. $\psi(\alpha_{\text{ref}})$ defines the feed-forward control loop in charge of computing the reference control torques corresponding to the angle-of-attack, provided by the guidance loop. The attitude control loop implemented in the HL-20 simulator is given in the Laplace domain as follows (for clarity, the Laplace variable s is omitted):

$$\begin{pmatrix} d_a \\ d_e \\ d_r \end{pmatrix} = K_2(\alpha_m)\bar{K}K_1(\alpha_m)\begin{pmatrix} \phi_{\text{ref}} - \phi_m \\ \alpha_{\text{ref}} - \alpha_m \\ \beta_{\text{ref}} - \beta_m \end{pmatrix} + \psi(\alpha_{ref}) - K_2(\alpha_m)\begin{pmatrix} p_m \\ q_m \\ r_m \end{pmatrix} \tag{7.54}$$

$$\delta = M(\bar{q}_m)(da, de, dr)^T. \tag{7.55}$$

7.7.1.2 Modeling the Dynamics

The dynamics of the HL-20 winged-body vehicle is derived from the classical 6 degree-of-freedom equations of motions of a symmetrical rigid aircraft. Under the assumptions of a flat and nonrotating earth, which is implicit with respect to the auto-landing trajectory duration, it follows that the dynamical behavior of the HL-20 winged-body vehicle can be described by the following nonlinear state representation:

$$\begin{cases} \dot{x}_b = f(x_b, \delta, \theta_p, u_w, v_w, w_w, C_x, C_y, C_z, C_l, C_m, C_n) \\ y_m = x_b + n \end{cases} \tag{7.56}$$

Table 7.1 HL-20 parameters variation ranges

Parameter	Variation range (%)	Dimension
M	$11{,}740 \pm 20$	[kg]
I_{xx}	$12{,}435 \pm 20$	[$Kg.m^2$]
I_{yy}	$67{,}716 \pm 20$	[$Kg.m^2$]
I_{zz}	$67{,}716 \pm 20$	[$Kg.m^2$]
C_{j0}	---- ± 20	[–]
X_{cg}	0.575 ± 20	% of the vehicle length

where $x_b = [u_b, v_b, w_b, \phi, \theta, p, q, r]^T$ denotes the state vector of the vehicle nonlinear model in the body frame. C_x, C_y, C_z, C_l, C_m, and C_n correspond to the dimensionless aerodynamic coefficients. u_w, v_w, and w_w represent the 3-axis wind and atmospheric turbulence components acting on the vehicle dynamics which are modeled by means of Dryden filters (see [53]). θ_p denotes a bounded parameters vector that models uncertainties inherent both to the vehicle design characteristics and to the aerodynamic coefficients, i.e.,

$$\begin{aligned} \theta_p = [&Ixx, Iyy, Izz, M, X_{cg}, Cx_0(\alpha,\beta), Cy_0(\beta), Cz_0(\alpha,\beta), \\ &Cl_0(\beta), Cm_0(\alpha,\beta), Cn_0(\alpha,\beta)]^T. \end{aligned} \tag{7.57}$$

Table 7.1 gives the variation ranges of M, I_{xx}, I_{yy}, I_{zz}, X_{cg}, C_{j0} for $j = \{x, y, z, l, m, n\}$.

To derive an analytical expression of (7.56), an important issue consists in obtaining an analytical expression of C_x, C_y, C_z, C_l, C_m, and C_n. The interested reader can refer to [54] where a solution based on principal component analysis and least-square polynomial interpolation techniques have been proposed. In the following, we assume that this modeling process has been completed and that an analytical expression of (7.56) is available.

The faults to be diagnosed correspond to any kind of faults occurring on the right and left wing flaps for which the remaining healthy control effectors are able to maintain the vehicle in-flight. Such faults can be modeled in a multiplicative manner according to (see Sects. 7.5 and 7.6 if necessary)

$$\begin{aligned} \delta_f &= (I_7 - \chi)\delta, \chi = diag(\chi_l, \chi_r, 0_5) \\ \delta &= [\delta_{wfl}, \delta_{wfr}, \delta_{bful}, \delta_{bfll}, \delta_{bfur}, \delta_{bflr}, \delta_r]^{\mathrm{T}} \end{aligned} \tag{7.58}$$

where χ_l, χ_r are unknown. The index "f" is used to outline the faulty cases. Note that $\chi_i = 1, i = \{l, r\}$ indicate that the left/right wing flap is out of order. Using the time dependency of $\chi_i = 1, i = \{l, r\}$, any type of faults can be considered, e.g., jamming, runaway, and oscillatory.

From Eqs. (7.56), (7.57), and (7.58), it follows that the overall HL-20 motion in faulty and fault-free situations can finally be written as follows:

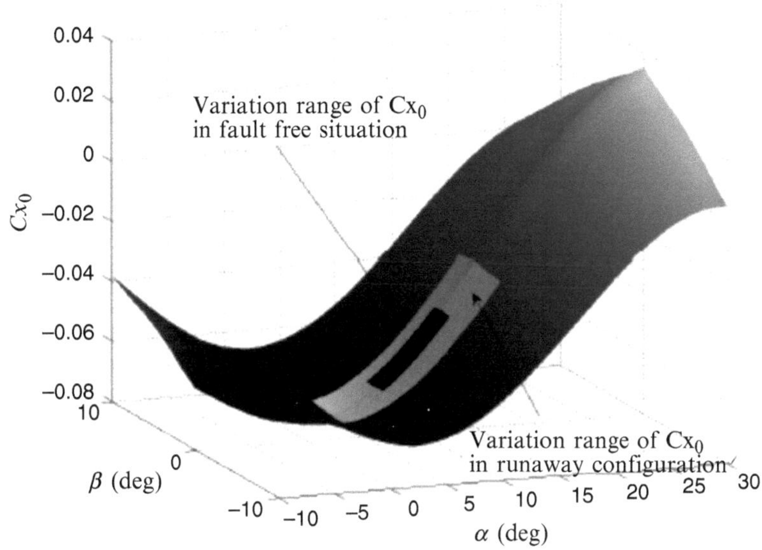

Fig. 7.20 $Cx_0(\alpha, \beta)$ coefficient for failed and fault-free configurations

$$\dot{x_b} = f(x_b, \delta_f, \theta_p, u_w, v_w, w_w). \tag{7.59}$$

$$\delta_f = (I_7 - \chi)\delta, \quad \chi = \text{diag}(\chi_l, \chi_r, 0_5). \tag{7.60}$$

$$y_m = x_b + n. \tag{7.61}$$

Remark 7.2 When an actuator failure occurs, both the guidance and attitude control loops will try to compensate the fault by means of the remaining control effectors. However, depending on the fault severity, the capability of the control loops to compensate faults may not be sufficient for some transient behavior. This leads $Cx_0(\alpha, \beta), Cy_0(\beta), Cz_0(\alpha, \beta), Cl_0(\beta), Cm_0(\alpha, \beta), Cn_0(\alpha, \beta)$ to vary outside their normal range of variations. Figure 7.20 shows these variation ranges for $Cx_0(\alpha, \beta)$, when a left wing flap actuator fault occurs. As it can be seen from the figure, the fault provokes a large change in the aerodynamics coefficients variation range as compared to the nominal case. It is then clear that the smaller the ability of the GNC to compensate the fault is, the larger this discrepancy will be. In other words, the fault models (7.58) may be inaccurate in some situations.

Using a first-order approximation of the nonlinear Eqs. (7.59), (7.60), and (7.61) around the flight path defined by $\phi_{\text{ref}}, \alpha_{\text{ref}}, \beta_{\text{ref}}$, one can derive an uncertain and time-varying linear state–space model of the HL-20 dynamics. To proceed, let $\xi x_b = x_b - x_{b_{\text{ref}}}$ be the tracking error between the current and reference states trajectories and $\xi\delta = \delta - \delta_{\text{ref}}$ the tracking error between the current and reference control signals.

Using a first-order Taylor expansion of Eqs. (7.59), (7.60), and (7.61), the dynamical error model describing the vehicle behavior around its flight trajectory is given by

$$\xi \dot{x_b} = A(x_{\text{bref}}, \delta_{\text{ref}}, \theta_p)\xi x_b + B(x_{\text{bref}}, \delta_{\text{ref}}, \theta_p)\xi\delta + E(x_{\text{bref}}, \delta_{\text{ref}}, \theta_p)w \tag{7.62}$$

$$\xi\delta_f = (I_7 - \chi)\xi\delta \tag{7.63}$$

$$\xi y_m = \xi x_b + n \tag{7.64}$$

where $A(x_{\text{bref}}, \delta_{\text{ref}}, \theta_p)$, $B(x_{\text{bref}}, \delta_{\text{ref}}, \theta_p)$ and $E(x_{\text{bref}}, \delta_{\text{ref}}, \theta_p)$ are matrices of appropriate dimensions parameterized by the reference states and controls. By expressing x_{bref} and δ_{ref} as functions of the reference flight velocity V_{ref}, it follows that the Eqs. (7.62), (7.63), and (7.64) can be re-written as follows:

$$\xi \dot{x_b} = A(V_{\text{ref}}, \theta_p)\xi x_b + B(V_{\text{ref}}, \theta_p)\xi\delta + E(V_{\text{ref}}, \theta_p)w. \tag{7.65}$$

$$\xi\delta_f = (I_7 - \chi)\xi\delta. \tag{7.66}$$

$$\xi y_m = \xi x_b + n. \tag{7.67}$$

Considering the slow variation of V_{ref} during the auto-landing phase (no abrupt acceleration or deceleration), it is assumed that V_{ref} can be fixed to a constant value. In other words, it is assumed that an adequate gridding by means of V_{ref} of the flight trajectory can be done, so that the aforementioned assumption yields. Thus, from (7.65), (7.66), and (7.67) and using an approximation of the actuator fault model (7.58) in terms of additive-type fault, it follows that the uncertain model of the HL-20 vehicle dynamics can be written according to the following LFT representation:

$$\xi y_m = F_u(P, \Delta)\begin{pmatrix} n \\ w \\ f \\ \xi\delta \end{pmatrix}. \tag{7.68}$$

$$\Delta = \text{blockdiag}(\delta_{c_x} I_{c_x}, \delta_{c_y} I_{c_y}, \delta_{c_z} I_{c_z}, \delta_m I_m, \delta_{I_{xx}} I_{I_{xx}}, \delta_{I_{yy}} I_{I_{yy}}, \delta_{I_{zz}} I_{I_{zz}}, \delta_{x_{cg}} I_{X_{cg}}),$$

$$\delta_k \in R : |\delta_k| \leq 1. \tag{7.69}$$

Noting now that α and $\bar{q}$ are close to constant values during the considered flight trajectory (auto-landing), the overall setup described in Fig. 7.19 can be modeled as illustrated in Fig. 7.21. This model is nothing else than the dynamics

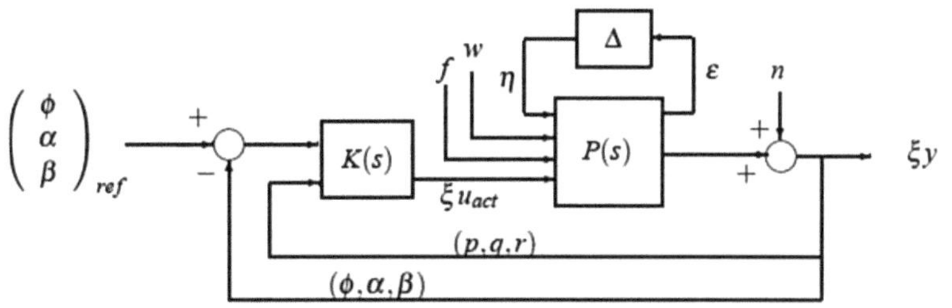

Fig. 7.21 HL-20 block diagram

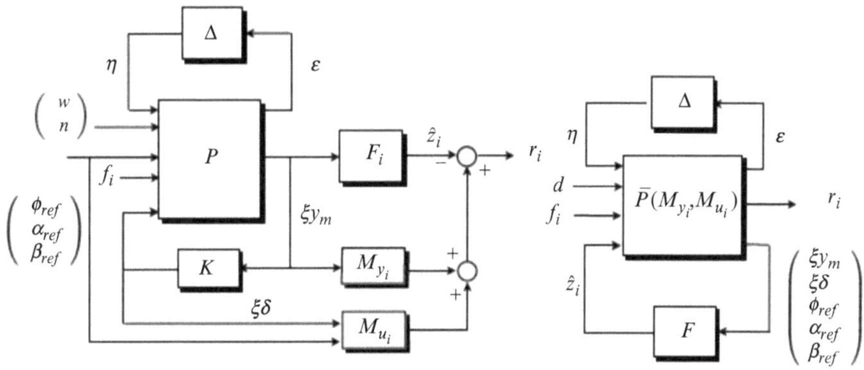

Fig. 7.22 Filter design problem formulation for wing flaps FDI

of the HL-20 model around the auto-landing trajectory defined by $\phi_{\rm ref}, \alpha_{\rm ref}, \beta_{\rm ref}$ taking into account the attitude control loop; the uncertainties I_{xx}, I_{yy}, I_{zz}, M, X_{cg}, $Cx_0(\alpha,\beta), Cy_0(\beta), Cz_0(\alpha,\beta), Cl_0(\beta), Cm_0(\alpha,\beta), Cn_0(\alpha,\beta)$; and the effect of actuator faults both on the vehicle and the GNC performances.

7.7.1.3 Design of the FDI System

The FDI strategy consists of a bank of two fault detection filters that are designed so that a given filter is made robust against measurement noise n, winds turbulences w, the guidance reference signals $\phi_{\rm ref}, \alpha_{\rm ref}, \beta_{\rm ref}$, and any fault in a given wing flap actuator while remaining sensitive to all faults in the others wing flap actuators.

The method used to develop this FDI unit involves two main steps:

Step 1: Firstly, two H_∞/H_- filters are designed using the model illustrated on Fig. 7.21 with a unique P evaluated at $V_{\rm ref} = 160$ m/s. The method is similar to the technique presented in Sect. 7.5 taking into account the uncertainty block Δ, i.e., the aim is to find two sets of matrices M_{yi}, M_{ui}, and filters $F_i(s)$, $i = 1, 2$ so that (see Fig. 7.22 for easy reference)

$$r_i = M_{yi}\xi y_m + M_{ui}\left(\xi\delta^{\mathrm{T}}, \phi_{\mathrm{ref}}, \alpha_{\mathrm{ref}}, \beta_{\mathrm{ref}}\right)^{\mathrm{T}} - F_i(s)\begin{pmatrix} \xi y_m \\ \left(\xi\delta^{\mathrm{T}}, \phi_{\mathrm{ref}}, \alpha_{\mathrm{ref}}, \beta_{\mathrm{ref}}\right)^{\mathrm{T}} \end{pmatrix},$$
$$i = 1, 2 \tag{7.70}$$

ξy_m and $\left(\xi\delta^{\mathrm{T}}, \phi_{\mathrm{ref}}, \alpha_{\mathrm{ref}}, \beta_{\mathrm{ref}}\right)^{\mathrm{T}}$ play the role of y and u in Sect. 7.5. Lemma 7.1 is then used to formulate the H_- specification as a fictitious H_∞ constraint, allowing thus the use of Proposition 7.1 for computing the structuring matrices M_{yi}, M_{ui}, and the filters $F_i(s)$, $i = 1, 2$. Thus and in order to avoid duplicating materials presented in Sect. 7.5, only the main steps of the design technique are presented in the following.

Step 2: Secondly, the capability of this unique FDI unit to fulfill the fault diagnosis task over all of the flight trajectory is tested using a μ_g analysis procedure to evaluate the conservativeness of the solution. The μ_g analysis procedure is a specially diagnosis-oriented tool for the assessment of robustness and sensitivity taking into account the nature (real or complex) and the structure (block diagonal) of the uncertainty block Δ [27, 28]. This aspect has not been considered in Sects. 7.5 and 7.6. So, the basic definitions for a good understanding of this analysis tool will be given below.

Definition of the μ_g Function

The μ framework considers perturbation blocks Δ satisfying a maximum norm constraint [55, 56]. In the μ_g framework, a second class of perturbations which satisfy a minimum gain constraint is considered. Consider a block structure $\underline{\Delta} = \mathrm{diag}(\underline{\Delta}_J, \underline{\Delta}_K)$ and a complex valued matrix $M = \begin{pmatrix} M_{JJ} & M_{JK} \\ M_{KJ} & M_{KK} \end{pmatrix}$ partitioned in accordance with $\underline{\Delta}$, which define the closed-loop equations

$$z = Mv, \quad v = \Delta z, \quad z = \left(z_j^{\mathrm{T}}\ z_k^T\right)^T, v = \left(v_j^T\ v_k^T\right)^T \tag{7.71}$$

where Δ_J and Δ_K satisfy, respectively, a maximum norm constraint and a minimum gain constraint. Then, the μ_g function is a positive real-valued function of the matrix M and the specified perturbation block $\underline{\Delta}$ defined by

$$\mu_{g\underline{\Delta}}(M) \triangleq \max_{\|v=1\|}\left\{\gamma : \begin{array}{l} \|v_j\|\,\gamma \le \|z_j\|, \forall j \in J \\ \|v_k\| \ge \|z_k\|\,\gamma, \forall k \in K \end{array}\right\}. \tag{7.72}$$

The μ_g function is defined on a domain $\mathrm{dom}(\mu_g)$ given by

$$M \in \mathrm{dom}(\mu g)\ \text{iff} \quad M_{KK}v_K = 0 \Rightarrow v_K = 0 \tag{7.73}$$

which is equivalent to a nontrivial solution, i.e., the maximization part in the μ_g problem is finite.

Fig. 7.23 Robust FDI performance problem

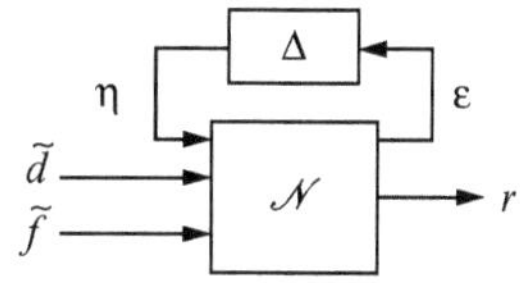

The μ_g Analysis Procedure

Consider the shaping filters W_d and W_f. Using some LFR algebra, it is always possible to derive from the H_∞/H_- design problem a LFR model according to the block diagram shown in Fig. 7.23 where $\tilde{d}$ is defined as $\tilde{d}(s) = W_d(s)d(s)$ and $\tilde{f}$ is defined according to $\tilde{f}(s) = W_f(s)f(s)$. In this formalism, all shaping filters and weighting functions (and the controllers) are included in $\mathcal{N}$.

The filter robust performance analysis problem over the model perturbation Δ is then a min–max gain problem over the specified frequency grid Ω, which can be formulated as follows:

$$\sup_{\omega} \bar{\sigma}\left(T_{\tilde{d}\to r}(j\omega)\right) < 1 \quad \text{and} \inf_{\omega\in\Omega} \underline{\sigma}\left(T_{\tilde{f}\to r}(j\omega)\right) > 1 \quad \forall \Delta \in \underline{\Delta} : ||\Delta||_\infty \leq 1. \tag{7.74}$$

In this equation, $T_{\tilde{d}\to r}$ and $T_{\tilde{f}\to r}$ denote respectively the closed-loop transfer between r and $\tilde{d}$ and r and f. The following theorem, which is an adaptation of the theorem 5 in [57], gives the solution of the robust fault sensitivity analysis problem within the μ_g framework. The proof is omitted here, as the main ideas can be found in [57].

Theorem 7.1 *Consider the model structure depicted in Fig.* 7.23 *and partition* $\mathcal{N}$ *according to* $\mathcal{N} = \begin{pmatrix} \mathcal{N}_{11} & \mathcal{N}_{12} \\ \mathcal{N}_{21} & \mathcal{N}_{22} \end{pmatrix}$ *where* $\mathcal{N}_{22}$ *denotes the transfer between the signals* r *and* $\tilde{f}$. *Let* $\sup_{\omega} \mu_{\bar{\underline{\Delta}}}(\mathcal{N}_{11}(j\omega)) < 1$ *with* $\bar{\underline{\Delta}} = \{\text{diag}(\Delta, \Delta_d)\}$ *where* $\Delta_d \in C^{\dim(\tilde{d})\times\dim(r)}$ *is a fictitious plant perturbation block introduced to close the loop between* r *and* $\tilde{d}$, *and let* $\mathcal{N} \in \text{dom}(\mu_g)$. *Then a necessary and sufficient condition for* (7.74) *to hold is*

$$\sup_{\omega\in\Omega} \mu_{g\hat{\underline{\Delta}}}(\mathcal{N}(j\omega)) < 1. \tag{7.75}$$

The block structure $\hat{\underline{\Delta}}$ *is defined according to* $\hat{\underline{\Delta}} = \left\{\text{diag}(\bar{\Delta}, \Delta_f)\right\}$. $\bar{\Delta}$ *is the perturbation block associated with the structure* $\bar{\underline{\Delta}}$ *defined above, and* $\Delta_f \in C^{\dim(\tilde{f})\times\dim(r)}$ *is a fictitious uncertainty block introduced to close the loop between* r *and* $\tilde{f}$.

Proof See [27].

The requirement $\sup_{\omega} \mu_{\bar{\underline{\Delta}}}(\mathcal{N}_{11}(j\omega)) < 1$ is equivalent to the maximum norm constraint in (7.74) is satisfied $\forall \Delta \in \underline{\Delta} : ||\Delta||_\infty \leq 1$, which is strictly equivalent to the robustness performance specification (S.1).

Since Theorem 7.1 is a necessary and sufficient condition which takes into account the structure and the nature of the model perturbations Δ, the robust sensitivity performance can be tested by calculating the μ_g function of $\mathcal{N}$ over the block structure $\hat{\underline{\Delta}}$ at those frequencies where the energy of the fault is likely to be concentrated. Furthermore, the gap between $\mu_{g\hat{\underline{\Delta}}}(\mathcal{N}(j\omega))$ and "1" provides a measure of the degree of conservatism of the FDI system. If for a particular frequency $\omega \in \Omega$ the computed μg is far from "1," then the fault sensitivity performance should be increased by reshaping the objective filters W_d and/or W_f.

This latest remark suggests that, from a practical point of view, Theorem 7.1 can be used within an iterative refinement procedure to derive the best achievable FDI performance:

1. Firstly, the filter state–space matrices A_F, B_F,C_F, D_F and the residual structuring matrices M_u, M_y are computed so that the H_∞/H_- requirements are met.
2. Secondly, according to Theorem 7.1, the robust performance is tested by evaluating the μ_g function. If the level of the achieved robustness and/or sensitivity performance is not judged satisfactory (i.e., if the μ and/or μ_g tests fail, respectively), go to step (*i*) and reshape the shaping filters W_d, W_f. New filter F and residual structuring matrices M_u, M_y are then designed. The procedure is stopped when best achievable performances are achieved.

Step 1: Design of M_{yi}, M_{ui}, $F_i\,(s)$, $i = 1, 2$

For the design task, since the major difficulty is concerned by the shaping filters W_d and W_f that allow us to specify the robustness and fault sensitivity requirements, only the guidelines to tune these dynamical filters are given here. The computation of M_{yi}, M_{ui}, $F_i\,(s)$, $i = 1, 2$ is also performed using Lemma 7.1 and Proposition 7.1 by means of the SDPT3 solver.

Since $d = \left(f_i \phi_{\text{ref}} \alpha_{\text{ref}} \beta_{\text{ref}} w^{\text{T}} n^{\text{T}}\right)^{\text{T}}$, it is natural to choose the shaping filter W_d according to

$$W_d = \text{diag}(W_{\bar{f}}, W_{\phi_{\text{ref}}}, W_{\alpha_{\text{ref}}}, W_{\beta_{\text{ref}}}, W_w, W_n). \tag{7.76}$$

$W_{\phi_{\text{ref}}}$, $W_{\alpha_{\text{ref}}}$, $W_{\beta_{\text{ref}}}$ enable to specify the robustness requirements against the guidance reference signals and W_w and W_n enable to formulate the robustness objectives against the atmospheric disturbances w (wind gust turbulences) and the measurement noise n. $W_{\bar{f}}$ is used to formulate the isolation objective.

Because we assume that the guidance signals may manifest themselves in a large frequency range (remember that they are generated through a guidance loop which depends on the reentry flight path), $W_{\phi_{\text{ref}}}$, $W_{\alpha_{\text{ref}}}$, $W_{\beta_{\text{ref}}}$ are fixed to a constant parameter γ_g. This means that it is required to reject the effects of the guidance signals on the residuals over all frequencies. With regard to W_w, since w is modeled as white noises through Dryden filters, $W_w = \text{diag}(W_{wu}, W_{wv}, W_{ww})$ is chosen as the inverse of the Dryden filters with static gain γ_w, i.e.,

$$W_{wu} = \gamma_w \left(\sigma_{ug} \sqrt{\frac{2L_{ug}}{\pi V_{\text{TAS}}}} \frac{1 + \tau s}{1 + \frac{L_{ug}}{V_{\text{TAS}}} s} \right)^{-1} \tag{7.77}$$

$$W_{wv} = \gamma_w \left(\sigma_{vg} \sqrt{\frac{2L_{vg}}{\pi V_{\text{TAS}}}} \frac{1 + \frac{2\sqrt{3}L_v}{V_{\text{TAS}}} s + \tau s}{\left(1 + \frac{L_{ug}}{V_{\text{TAS}}} s\right)^2} \right)^{-1} \tag{7.78}$$

$$W_{ww} = W_{wv}. \tag{7.79}$$

V_{TAS} denotes the vehicle flight velocity (in m/s), L_u, L_v, L_w are the turbulence scale lengths, and $\sigma_u, \sigma_v, \sigma_w$ the turbulences intensities. L_u, L_v, L_w and $\sigma_u, \sigma_v, \sigma_w$ being nonlinear functions of altitude and velocity, the retained numerical values of W_{wu},W_{wv},W_{ww} considered in the design procedure correspond to moderate turbulence flight conditions for altitude higher than 1,000 ft. τ is a high-frequency zero introduced to make W_{wu},W_{wv}, and W_{ww} invertible. By this choice, it is desired to have a rejecting behavior of $T_{wr}\ (j\omega)$ at those frequencies where w manifests itself. Here $T_{wr}\ (s)$ denotes the transfer between the residual r and w.

The energy content of the measurement noise is assumed to be located in the frequency range $[100, +\infty]$ rad/s, so W_n is fixed as a low-pass filter with static gain γ_n, i.e.,

$$W_n = \gamma_n \frac{1 + \frac{1}{100}s}{1 + \frac{1}{5}s} . I_{10}. \tag{7.80}$$

In others words, it is desired to have a rejecting behavior of $T_{nr}\ (j\omega)$ at the frequencies $[100, +\infty]$ rad/s. T_{nr} is used to denote the transfer between r and n. Note that the parameters γ_g, γ_w, and γ_n have been introduced in order to manage the robustness level of the fault detection scheme against $\phi_{ref}, \alpha_{ref}, \beta_{ref}$, w, and n separately.

For isolation and fault sensitivity, we assume that all faults (e.g., locked-in-place, runaway, floating surface, loss-of-effectiveness) have low-frequency behavior. So W_f is taken as a low-pass filter with cutoff frequency ω_f

$$W_f = \gamma_{2_i} \frac{1}{1 + \tau_f s}, \quad \tau_f = \frac{1}{\omega_f} \tag{7.81}$$

$$W_{\bar{f}}(s) = \gamma_{\bar{f_i}} W_f^{-1}(s). \tag{7.82}$$

The parameters γ_{2i} and $\gamma_{\overline{f_i}}$ have been introduced to manage the sensitivity and isolation performances separately.

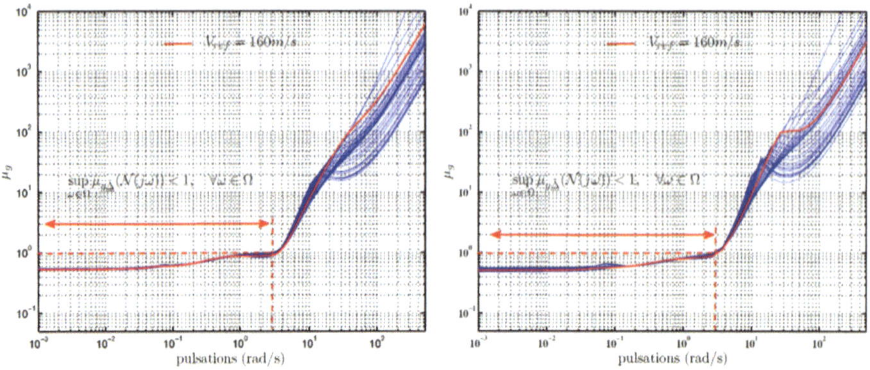

Fig. 7.24 Frequency behavior of $\mu_g(\mathcal{N}(j\omega))$ for (M_{y_1}, M_{u_1}, F_1) (*left*) – (M_{y_2}, M_{u_2}, F_2) (*right*)

The parameters $\gamma_n, \gamma_w, \gamma_g, \gamma_{2i}, \gamma_{\bar{f}i}, \omega_{fi}$, $i = 1, 2$ are then optimized through the previously explained iterative refinement following the design/μg analysis cycle.

Step 2: μ_g Analysis Procedure

Because the two diagnosis filters have been designed on the basis on a unique $P(s)$ evaluated at $V_{\text{ref}} = 160$ m/s (see (7.68)) (remember that one filter is dedicated to the left wing flap and that the other is dedicated to the right wing flap), the μ_g analysis procedure is now used to test the ability of the FDI system to diagnose the considered faults on the overall flight trajectory. Note that the design of M_{yi}, M_{ui}, and F_i (s), $i = 1, 2$ involves sufficient conditions and does not take into account the structure (block diagonal) and the nature (real or complex) of the model perturbation block Δ. Thus, the designed FDI filters may be conservative. To check if the required FDI objectives are achieved over the model perturbations Δ, Theorem 7.1 is considered. Note that this analysis procedure has not been considered in Sects. 7.5 and 7.6 since there did not exist uncertainties Δ for the MICROSCOPE case and since the considered technique for the MSR application involves H (0)/RLMI-based robust pole assignment technique and thus differ from the H_∞/H_- framework.

To proceed, the flight trajectory is parameterized using the velocity V_{ref} and a gridding of this latter is performed every 2.5 m/s; 30 flight points are considered. For each velocity, the corresponding $P(s)$ is computed and the μ_g setup illustrated in Fig. 7.23 is synthesized taking into account the FDI filters (M_{yi}, M_{ui}, F_i (s)), $i = 1, 2$. The μ_g analysis procedure is finally performed using Theorem 7.1, taking into account the uncertainty block Δ. Figure 7.24 illustrates the behavior of the μ_g function over the frequency range [0; 500] rad/s for the considered velocity V_{ref}.

As it can be seen, $\mu_{g\underline{\tilde{\Delta}}}(\mathcal{N}(j\omega)) < 1$ over the frequency range $\Omega \approx [0, 2.5]$ rad/s for both filters which indicate that the required robustness and fault sensitivity are achieved for all specified uncertainties (see Table 7.1) along the flight trajectory. Furthermore, the small gap between μ_g and 1 indicates that

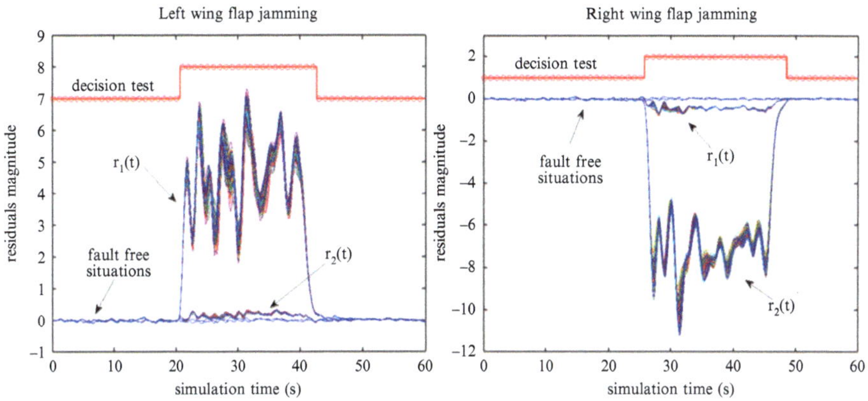

Fig. 7.25 Left wing flap jamming (*left*) – right wing flap jamming (*right*)

the solution is not too conservative. Finally the frequency range $\Omega \approx [0, 2.5]$ rad/s where $\mu_{g\tilde{\underline{\Delta}}}(\mathcal{N}(j\omega)) < 1$ indicates that the computed FDI exhibits short transient responses (about $1/2.5 = 0.4\,s$) while maintaining both robustness and fault sensitivity performance despite the variations of I_{xx}, I_{yy}, I_{zz}, M, X_{cg}, and $Cx_0(\alpha,\beta), Cy_0(\beta), Cz_0(\alpha,\beta), Cl_0(\beta), Cm_0(\alpha,\beta), Cn_0(\alpha,\beta)$.

7.7.2 Nonlinear Simulations

The FDI system is next implemented within the HL-20 nonlinear simulator. All simulations are run during the auto-landing phase (see Fig. 7.18). Different fault types are introduced on the right and left wing flap actuators. Ten faulty scenarios are considered. However, for brevity, only a few results are presented in this chapter:

- The first faulty scenario corresponds to a jamming (locked-in-place). At $t = 20$ s, the left wing flap is locked at its current value, i.e., $\delta_{wfl} = 12°$, and is unblocked from $t = 40$ s until the end of the simulation.
- The second faulty scenario is the right wing flap jamming at $\delta_{wfr} = 9.7°$ on the time interval $25\text{ s} < t < 45$ s.
- The third one is a fast motion to its extreme position (i.e., 30°) of the right wing flap, i.e., runaway-type fault. This fault occurs during $25\text{ s} < t < 40$ s.
- The fourth simulated scenario is a runaway-type fault occurring in the right wing flap between $t = 30$ s and $t = 45$ s.

Figures 7.25 and 7.26 illustrate the residuals behavior $r_i(t)$, $i = 1,2$ in fault-free and faulty situations. 1,000 nonlinear simulations have been run with different combinations of uncertainties (see Table 7.1). As expected, $r_1(t)$ (r_2(t), respectively)

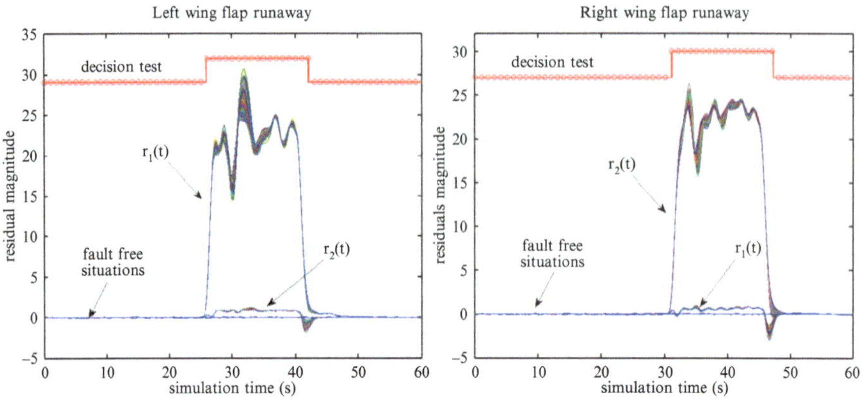

Fig. 7.26 Left wing flap runaway (*left*) – right wing flap runaway (*right*)

is sensitive (in the H_--norm sense) to any type of faults occurring in the left (the right, respectively) wing flap actuator while remaining robust (in the H_∞-norm sense) against faults occurring in the right (the left, respectively) wing flap, wind turbulences, measurement noises, whatever the fluctuated guidance signals generated by the guidance loop to compensate faults, and the parametric uncertainties.

Clearly the nonlinear simulations show that faults are shortly detected and isolated by the proposed FDI unit.

7.7.3 Application to the TAEM Phase

The same design procedure is now applied for actuator fault diagnosis during the TAEM phase, i.e., $0.5 \leq \text{Mach} \leq 2$. The goal is to demonstrate how the H_∞/H_- approach can handle such a highly nonlinear flying phase.

Here, the major differences with the auto-landing phase are:

- The aerodynamic coefficients: Due to the hypersonic ($\text{Mach} > 1$) and transonic (gate at $\text{Mach} = 1$) phases, the aerodynamic coefficients are highly nonlinear functions of flying conditions such as the angle-of-attack and the Mach number.
- The GNC system dedicated to the TAEM phase which is different than that for the auto-landing phase.

The aerodynamic coefficients are given in the Mach-α map according to the following equations[7]:

[7] The complete database of the aerodynamic coefficients is available at http://ntrs.nasa.gov

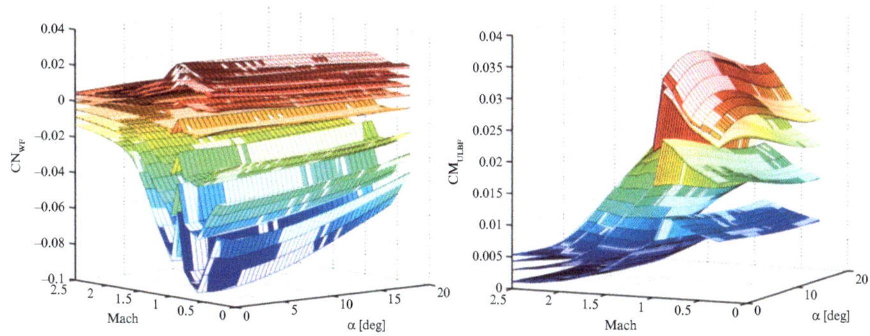

Fig. 7.27 Identified aerodynamic coefficients $C_{n_{wfl}}$ (*left*) and $C_{n_{wfl}}$ (*right*) superimposed with their database

$$
\begin{bmatrix} C_L \\ C_Y \\ C_D \\ C_l \\ C_m \\ C_n \end{bmatrix} = \begin{bmatrix} C_{L_0}(\alpha, \text{Mach}) \\ C_{Y_0}(\alpha, \text{Mach})\beta \\ C_{D_0}(\alpha, \text{Mach}) \\ C_{l_0}(\alpha, \text{Mach})\beta \\ C_{M_0}(\alpha, \text{Mach}) \\ C_{N_0}(\alpha, \text{Mach})\beta \end{bmatrix} + \sum_j \begin{bmatrix} C_{Lj}(\alpha, M, \delta_j) \\ C_{Yj}(\alpha, M, \delta_j) \\ C_{Dj}(\alpha, M, \delta_j) \\ C_{lj}(\alpha, M, \delta_j) \\ C_{mj}(\alpha, M, \delta_j) \\ C_{nj}(\alpha, M, \delta_j) \end{bmatrix}
$$

$$
+ \frac{1}{2V} \begin{bmatrix} 0 \\ 0 \\ 0 \\ b.(C_{lp}(\alpha).p + C_{lr}(\alpha).r) \\ C_{lq}(\alpha).q.\bar{c} \\ b.(C_{np}(\alpha).p + C_{nr}(\alpha).r) \end{bmatrix} \begin{bmatrix} C_x \\ C_y \\ C_z \end{bmatrix} \tag{7.83}
$$

$$
= R_{bs} \begin{bmatrix} -C_L \\ C_Y \\ -C_D \end{bmatrix}, j = \{\delta_{wfl}, \delta_{wfr}, \delta_{urbf}, \delta_{ulbf}, \delta_{lrbf}, \delta_{llbf}, \delta_r\}
$$

where C_x, C_y, C_z, C_l, C_m, and C_n correspond to the dimensionless aerodynamic coefficients. R_{bs} denotes the body-stability matrix rotation.

As all terms in this equation are stored in numerical look-up-tables, it is necessary to establish from the aerodynamic coefficients database, an analytical expression of the terms C_{L0}, C_{Y0},C_{D0},C_{l0}, C_{m0},C_{n0}, C_{Lj}, C_{Yj}, C_{Dj}, C_{lj}, C_{mj}, C_{nj}, C_{lp}, C_{lr}, C_{lq}, C_{np}, and C_{nr}. Sigmoid-based neural networks are used for this purpose. Remember that the establishment of the LFR model illustrated in Fig. 7.21 is based on a first-order Taylor series expansion of Eqs. (7.59), (7.60), and (7.61) around its reference flight trajectory. Figure 7.27 illustrates the obtained results for the aerodynamic coefficients of the left wing flap $C_{n_{wfl}}(\alpha, \text{Mach}, \delta_{wfl})$ and the upper left body-flap $C_{M_{ulbf}}(\alpha, \text{Mach}, \delta_{ulbf})$ for some control surface position δ_{wfl} and δ_{ulbf}.

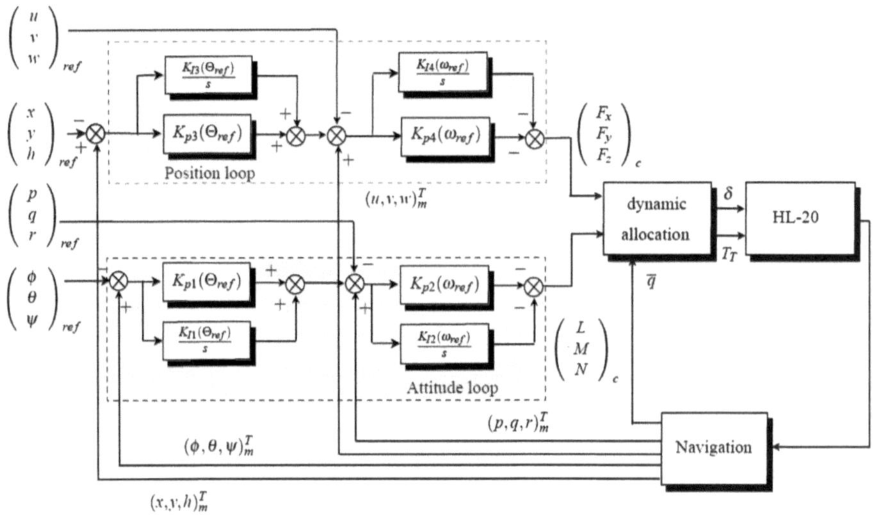

Fig. 7.28 GNC architecture for the TAEM phase of HL-20

As it can be seen, the estimated models successfully approximate the numerical values of the associated aerodynamic coefficients stored in the database. Note that these plots clearly illustrate the nonlinearities of the aerodynamic coefficients during the transonic phase (Mach = 1).

The GNC system used for the TAEM phase consists of two main loops that are dedicated to the attitude and the position of the vehicle. The attitude loop is in charge of generating the moments L, M, N, and the position loop is in charge of the forces F_x, F_y, F_z. Forces and moments are then converted into control surfaces angles (and thrusts T_T for the remote control system if required) by means of a dynamic control allocation scheme depending on the dynamic pressure $\bar{q}$. The overall architecture of the GNC for the TAEM phase is given in Fig. 7.28.

The scheduled gains K_{I1}, K_{I2}, K_{I3}, K_{I4} and K_{p1}, K_p, K_{p3}, $Kp4$ that depend on the required reentry trajectory parameters $\omega_{\text{ref}} = (pqr)^{\mathrm{T}}_{\text{ref}}$ and $\Theta_{\text{ref}} = (\phi\theta\psi)^{\mathrm{T}}_{\text{ref}}$ have been determined by means of a poles assignment technique in order to satisfy the control objectives. Note that the control loops are PI-based controllers.

Similarly to the procedure used for the auto-landing phase, a H_∞/H_- fault detector is considered for wing flaps fault detection. The strategy consists of a H_∞/H_- filter that is designed based on a unique model of the HL-20 evaluated at a fixed reference speed $V_{\text{ref}} = 400$ m/s. The aim is to find the structuring matrices M_y, M_u and the filter $F(s)$ so that

$$r = M_y \xi y_m + M_u \xi \delta - F(s) \begin{pmatrix} \xi y_m \\ \xi \delta \end{pmatrix}. \tag{7.84}$$

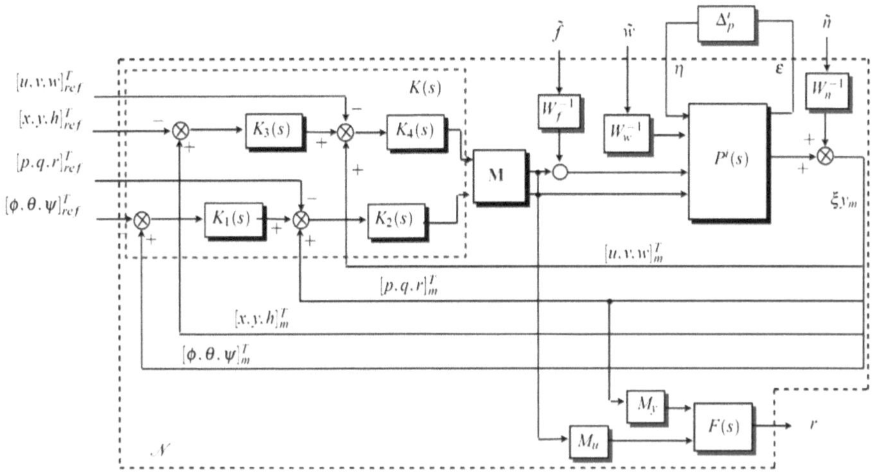

Fig. 7.29 H_∞/H_- setup for the TAEM phase

Secondly, the capability of this unique FDI unit to fulfill the fault detection task over the whole TAEM trajectory is analyzed using the μ_g analysis procedure. A gridding of 80 numerical values for V_{ref} has been considered for this purpose, i.e., the FDI scheme unit is analyzed for 200 m/s $\leq V_{\text{ref}} \leq$ 600 m/s (0.5 $\leq$ Mach $\leq$ 2) with a step equal to 5 m/s. The conservativeness of the solution is evaluated, again, by means of the μ_g analysis procedure.

Figure 7.29 illustrates the general diagram that is used for the H_∞/H_- design and the μ_g analysis of the FDI system.

In Fig. 7.29:

- $K_i^j(s), i = 1,\ldots,4,\ j = 1,\ldots,80$ denote the model of the GNC for a fixed $V_{\text{ref.}}$
- $\mathbf{M}^j, j = 1, \ldots, 80$ denote the model of the allocation unit for a fixed ^{V}ref.
- $W_f, W_w^j, W_n,\ j = 1,\ldots,80$ are the shaping filters specifying the fault sensitivity and robustness objectives of the residual generator (7.84). These shaping filters are chosen to be the same than those used for the auto-landing phase; see the above section. For Dryden filters, it is considered numerical values for altitude higher than 2,000 ft.
- $F_u(P^j, \Delta^j),\ j = 1,\ldots,80$ denote the LFR model of the HL-20 dynamics for a given and fixed V_{ref}. Each LFR has been obtained using a first-order approximation of the Eq. (7.56) around the TAEM trajectory. In terms of fault modeling, the model (7.58) is used. The considered uncertainties are those listed in Table 7.1. These considerations boil down to the following definition of the LFR $F_u(P^j, \Delta^j)$:

$$\Delta^j = \text{diag}\left\{\delta_{C_{l0}}, \delta_{C_{m0}}, \delta_{C_{n0}}, \delta_{C_{x0}} I_3, \delta_{X_{cg}} I_2, \delta_{C_{y0}} I_3, \delta_{C_{z0}} I_3, \ldots\right.$$
$$\left.\delta_{I_{xx}} I_2, \delta_{I_{yy}} I_3, \delta_{I_{zz}} I_3, \delta_m I_3\right\}, \quad \dim(\Delta^j) = 27, j = 1,\ldots,80. \qquad (7.85)$$

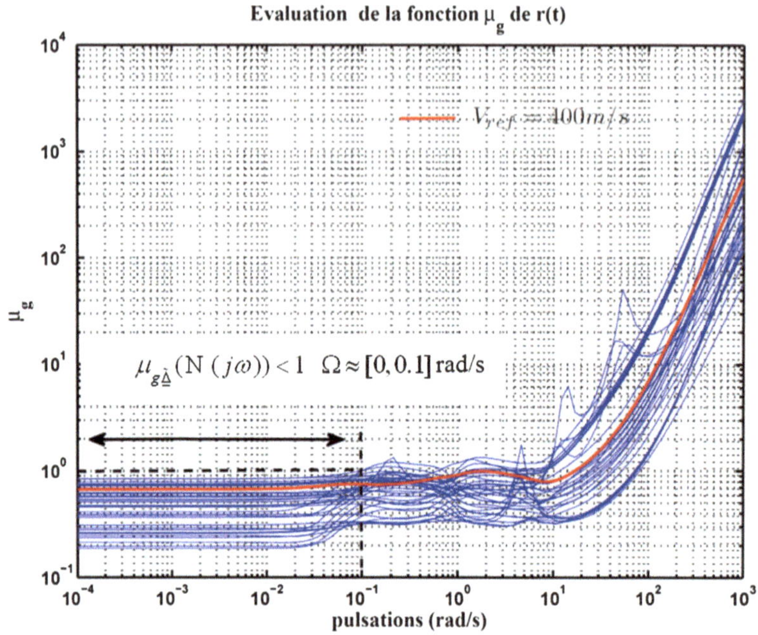

Fig. 7.30 Behavior of the μ_g function – TAEM phase

Figure 7.30 illustrates the behavior of the μg function over the frequency range [0; 1,000] rad/s for 200 m/s $\leq V\text{ref} \leq$ 600 m/s (0.5 $\leq$ Mach $\leq$ 2). The case $V\text{ref} = 400$ m/s is outlined in red to better appreciate the μg analysis corresponding to the model used for the H_∞/H_- design step. As it can be seen, $\mu_{g\tilde{\underline{\Delta}}}(\mathcal{N}(j\omega)) < 1$ in the frequency range $\Omega \approx [0, 0.1]$ rad/s for any Vref. Following Theorem 7.1, this indicates that both the required robustness against the measurement noise n, the winds turbulences w and the guidance reference signals, and the fault sensitivity performance are achieved for all considered parametric uncertainties (see Table 7.1) and along the entire flight trajectory, i.e., for 0.5 $\leq$ Mach $\leq$ 2. Furthermore, it can be noticed that for many reference speeds V_{ref}, $\mu_{g\tilde{\underline{\Delta}}}(\mathcal{N}(j\omega)) < 1$ over the frequency range $\Omega \approx [0, 10]$ rad/s, showing first, that the robustness and fault sensitivity specifications are met for a large range of values of V_{ref} and, second, that for those values, the performances are guaranteed over a larger frequency range (two decades more). Finally, the small gap between μ_g and 1 indicates that the computed solution (M_y, M_u, $F(s)$) is not so conservative.

7.7.4 Nonlinear Simulations

The fault detection filter is next converted to discrete time using a Tustin approximation and implemented within the nonlinear simulator of the HL-20. Different

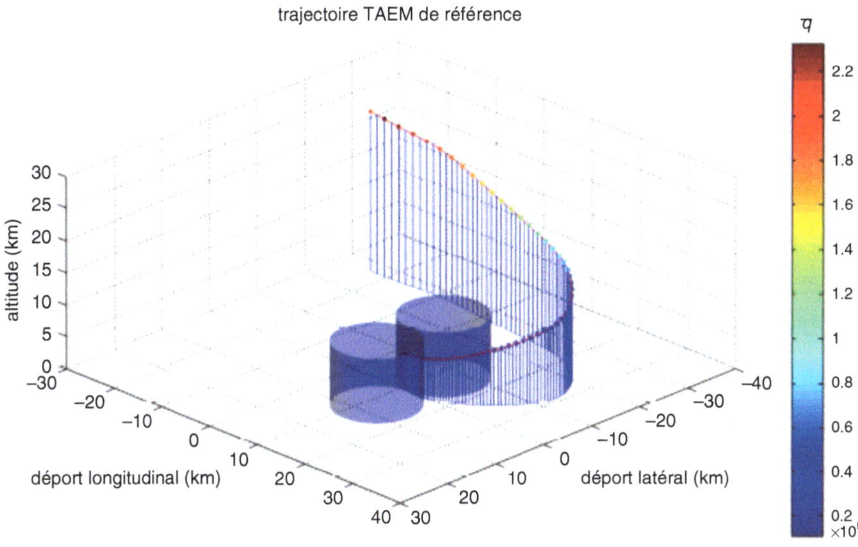

Fig. 7.31 TAEM trajectory

faulty scenarios are considered. The time occurrence of the fault is chosen arbitrary so that all phases of the TAEM trajectory are considered, i.e., the supersonic ($1 < \text{Mach} \leq 2$), the transonic ($\text{Mach} = 1$), and the subsonic phases ($\text{Mach} < 1$). The time occurrence of the gate $\text{Mach} = 1$ is indicated on the figures. The faulty scenarios correspond to (from top left to right bottom):

- The left wing flap is jammed at null position during the flight at $t = 150$ s and remains fault-free at $t = 200$ s. The fault occurs after the transonic ($\text{Mach} = 1$) phase.
- The right wing flap is jammed at null position during the flight for $100\text{ s} \leq t \leq 150\text{ s}$ (the transonic phase occurs during this time).
- A runaway at maximum deflection speed occurs in the left wing flap for $150\text{ s} \leq t \leq 220\text{ s}$. The fault occurs after the transonic phase.
- A runaway at maximum deflection speed occurs in the right wing flap before the transonic phase and is maintained after it.
- A loss of efficiency (gain variation) of the left wing flap occurs after the transonic phase and is maintained until $t = 170$ s.
- The last scenario is the same as the previous one with respect to the right wing flap.

The TAEM reentry trajectory used for the simulations is presented in Fig. 7.31. The dynamic pressure $\bar{q}$ is presented too, to better appreciate the characteristics of the considered trajectory.

Figures 7.32 and 7.33 illustrate the behavior of the residual $r(t)$ and the decision-making procedure. As expected, $r(t)$ is sensitive (in the H_--norm sense) to any

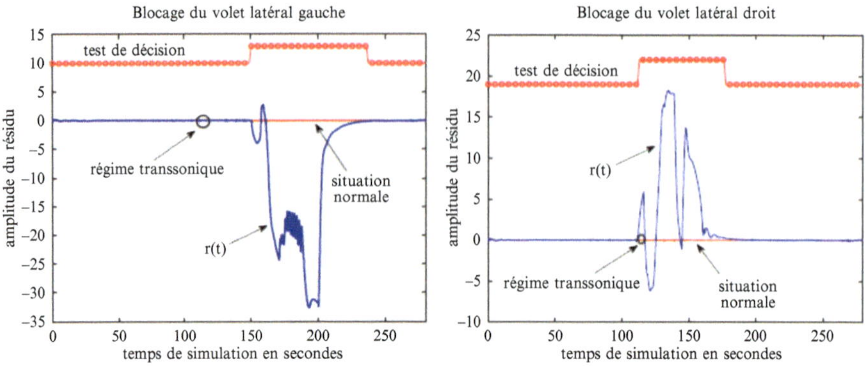

Fig. 7.32 Behavior of *r*(*t*) and decision-making test. Jamming of the left (*left*) and right (*right*) wing flap

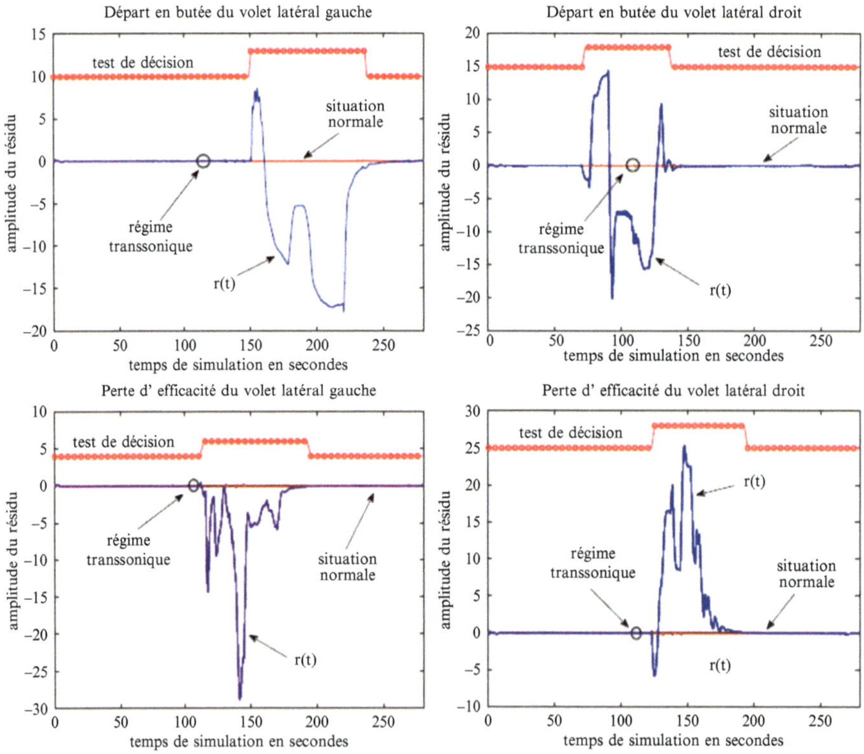

Fig. 7.33 Behavior of *r*(*t*) and decision-making test. From *top left* to *bottom right*: runaway at maximum deflection speed and loss of efficiency of the left and right wing flap

type of faults occurring in the left (the right, respectively) wing flap actuator while remaining robust (in the H_∞-norm sense) against wind turbulences, measurement noises, guidance signals, and uncertainties (see Table 7.1). Clearly the nonlinear simulations show that faults are shortly detected by the proposed fault detector despite the nonlinearities involved in the aerodynamic coefficients.

7.8 Conclusion

This chapter investigated the problem of fault detection and diagnosis for space missions. Three application cases have been studied: an Earth observation satellite, a deep space mission, and an atmospheric reentry vehicle. In each case, the focus has been first on careful modeling which is of primary importance for successful FDI of such applications. The effect of Guidance, Navigation, and Control has been analyzed and taken into account. Modeling stage should also take into account uncertainties stemming from a large variety of spatial disturbances and endogenous sources and their propagation. The FDI design/analysis method presented in this chapter is quite general and takes advantage of H_∞/H_-, generalized μ analysis, and powerful numerical LMI tools. The approach provides a nice iterative refinement procedure which allows the designer to get a good balance between various robustness/performance specifications and trade-offs.

References

1. Bornschleg E (2008) Fdir requirements and rational – esa r&d activities overview for gnc and software. In: CCT CNES, Toulouse, France
2. Olive X (2010) Fdi(r) for satellite at thales alenia space: how to deal with high availability and robustness in space domain? In: Conference on control and fault-tolerant systems (SysTol'10). IEEE, Nice, France, pp 837–842
3. Henry D, Simani S, Patton R (2010) Fault detection and diagnosis for aeronautic and aerospace missions in 'fault tolerant flight control: a benchmark challenge'. Springer, Berlin
4. Satin A, Gates R (2005) Evaluation of parity equations for gyro failure detection and isolation. J Guid Control 1(1):14–20
5. Shim DS, Yang CK (2004) Geometric FDI based on SVD for redundant inertial sensor systems. In: 2004 5th Asian control conference, IEEE, vol. 2, Melbourne, Victoria, Australia, pp 1094–1100
6. Yang C-K, Shim D-S (2007) Double faults isolation based on the reduced-order parity vectors in redundant sensor configuration. Int J Control Autom Syst 5(2):155–160
7. Verma V, Langford J, Simmons R (2001) Non-parametric fault identification for space rovers. In: International Symposium on Artificial Intelligence and Robotics in Space (iSAIRAS). J.-C. Piedboeuf, June 2001
8. DeFreitas N (2002) Rao-blackwellised particle filtering for fault diagnosis. In: Aerospace conference, IEEE, vol. 4, pp 1767–1772
9. Hutter F, Dearden R (2003) Efficient on-line fault diagnosis for non-linear systems. In: International Symposium on Artificial Intelligence, Robotics and Automation in Space (iSAIRAS), Nara, Japan, 19–23 May 2003

10. Venkateswaran N, Siva M, Goel P (2002) Analytical redundancy based fault detection of gyroscopes in spacecraft applications. ACTA Astron 50(9):535–545
11. Jensen H, Wisniewski R (2002) Fault detection and isolation for spacecraft: geo- metric approach. In: Guidance navigation and control. AIAA, Monterey, CA, 5–8 Aug 2002
12. Edwards C, Spurgeon S, Patton R (2000) Sliding mode observers for fault detection and isolation. Automatica 36:541–553
13. Tan C, Edwards C (2003) Sliding mode observers for robust detection and re- construction of actuator and sensor faults. Int J Robust Nonlinear Control 13:443–446
14. Edwards C, Thein M (2006) Sliding mode fault detection and isolation in a satellite leader/follower system. In: Proceedings of SAFEPROCESS'2006. IFAC, Beijing, China, pp 367–372
15. Patton R, Uppal F, Simani S, Polle B (2006) A monte carlo analysis and design for fdi of a satellite attitude control system. In: Proceedings of SAFEPROCESS'2006. IFAC, Beijing, China, pp 1393–1398
16. Dearden R, Clancy D (2002) Particle filters for real-time fault detection in planetary rovers. In: Proceedings of the thirteenth international workshop on principles of diagnosis, pp 1–6
17. Dearden R, Willeke T, Simmons R, Verma V, Hutter F, Thrun S (2004) Real- time fault detection and situational awareness for rovers: report on the mars technology program task. In: Aerospace, IEEE, 6–13 Mar 2004, pp 826–840
18. Patton R, Uppal F, Simani S, Polle B (2008) Reliable fault diagnosis scheme for a spacecraft attitude control system. Proc IMechE Part 0: J Risk Reliab 222:139–152
19. Patton R, Uppal F, Simani S, Polle B (2010) Robust fdi applied to thruster faults of a satellite system. Control Eng Pract 18(9):1093–1109
20. Alwi CEH, Marcos A (2010) Fdi for a mars orbiting satellite based on a sliding mode observer scheme. In: Conference on control and fault-tolerant systems (SysTol). IEEE, Nice, France, pp 125–130
21. Castro H, Bennani S, Marcos A (2006) Robust filter design for a re-entry vehicle. In: Proceedings of the 7th international conference on dynamics and control of systems and structures in space, Greenwish, UK
22. Murray K, Marcos A, Nin LP, Bornschlegl E (2008) Gain scheduled fdi for a re-entry vehicle. In: AIAA guidance, navigation and control conference and exhibit. AIAA, Honolulu, Hawaii
23. Falcoz A, Henry D, Zolghadri A, Bornschleg E, Ganet M (2008) On-board model-based robust fdir strategy for reusable launch vehicles (rlv). In: 7th international ESA conference on guidance, navigation and control systems, County Kerry, Ireland
24. Falcoz A, Henry D, Zolghadri A (2010) Robust fault diagnosis for atmospheric re-entry vehicles: a case study. IEEE Trans Syst Man Cybern – Part A: Syst Hum 40(5):886–899
25. Falcoz A, Boquet F, Dinh M, Polle B, Flandin G, Bornschlegl E (2010) Robust fault diagnosis for spacecraft: application to lisa pathfinder experiment. In: Conference on control and fault-tolerant systems (ACA'2010). IFAC, Nara, Japan
26. Falcoz A, Boquet F, Flandin G (2010) Robust h/h_{-} thruster failure detection and isolation with application to the lisa pathfinder spacecraft. In: AIAA guidance, navigation, and control conference, AIAA, Toronto, Ontario
27. Henry D, Zolghadri A (2005) Design and analysis of robust residual generators for systems under feedback control. Automatica 41:251–264
28. Henry D, Zolghadri A (2005) Design of fault diagnosis filters: a multi-objective approach. J Franklin Inst 342(4):421–446
29. Henry D, Zolghadri A, Monsion M, Ygorra S (2002) Off-line robust fault diagnosis using the generalised structured singular value. Automatica 38(8):1347–1358
30. Patton R, Frank P, Clark R (2000) Issues of fault diagnosis for dynamic systems. Springer, London. ISBN 3-540-19968-3
31. Chen J, Patton R (1999) Robust model-based fault diagnosis for dynamic systems. Kluwer Academic, Boston
32. Beck C, Doyle J, Glover K (1996) Model reduction of multidimensional and uncertain systems. IEEE Trans Autom Control 41(10):1466–1477

33. Cockburn J, Morton B (1997) Linear fractional representations of uncertain systems. Automatica 33(7):1263–1271
34. Varga A, Looye G, Moormann D, Grübel G (1998) Automated generation of LFT-based parametric uncertainty descriptions from generic aircraft models. Math Model Dyn Syst 4: 249–274
35. Beck C, Doyle J (1999) A necessary and sufficient minimality condition for uncertain systems. IEEE Trans Autom Control 44(10):1802–1813
36. Hecker S, Varga A, Magni J (2005) Enhanced LFR-toolbox for matlab. Aerosp Sci Technol 9:173–180
37. Wied B (1998) Space vehicle dynamics and control. American Institute of Aeronautics and Astronautics, Reston
38. Wisniewski R (2000) Lecture notes on modelling of a spacecraft. Aalborg Universitet, Aldeling for Proceskontrol, Aalborg
39. Jin H, Wiktor P, DeBra D (1995) An optimal thruster configuration design and evaluation for quick step. Control Eng Pract 3(8):1113–1118
40. Wiktor P (1996) On-orbit thruster calibration. J Guid Control Dyn 19(4):934–940
41. Henry D (2008) Fault diagnosis of the MICROSCOPE satellite actuators using H/H_- filters. AIAA J Guid Control Dyn 31(3):699–711
42. Isermann R (2005) Model-based fault detection and diagnosis – status and applications. Annu Rev Control 29:71–85
43. Zolghadri A, Goetz C, Bergeon B, Denoize X (1998) Integrity monitoring of flight parameters using analytical redundancy. In: IEEE international conference on Control'98, Swansea, UK
44. Das I, Dennis J (1997) A closer look at drawbacks of minimizing weighted sums of objective for pareto set generation in multicriteria optimization problems. Struct Optim 14(1):63–69
45. Henry D, Zolghadri A (2006) Norm-based design of robust fdi schemes for uncertain systems under feedback control: comparison of two approaches. Control Eng Pract 14(9):1081–1097
46. Gahinet P, Apkarian P (1994) A linear matrix inequality approach to H control. Int J Robust Nonlinear Control 4:421–428
47. CNES (1998) Cours de technologie spatiale, Techniques et Technologies des ve´ hicules spatiaux, vol 1,2,3. CEPADUES-ED
48. Wertz JR, Larson WJ (1999) Space mission analysis and design, 3rd edn. Springer, New York
49. Irvin DJ (2001) A study of linear vs. nonlinear control techniques for the reconfiguration of satellite formations. PhD thesis, Department of Aeronautics and Astronautics – Graduate School of Engineering and Management – Air Force Institute of Technology – Air University – Air Education and Training Command, Wright-Patterson Air Force Base, Ohio
50. Zames G (1966) On the input-output stability of time varying nonlinear feedback systems. In: IEEE transactions on automatic control, Cambridge, MA, pp 228–238, 465, 467
51. Scherer C, Gahinet P, Chilali M (1997) Multiobjective output-feedback control via LMI optimization. IEEE Trans Autom Control 42(7):896–911
52. Chilali M, Gahinet P, Apkarian P (1999) Robust pole placement in LMI regions. IEEE Trans Autom Control 44(12):2257–2270
53. der Linden CV (1996) DASMAT-Delft University aircraft simulation model and analysis tool. Faculty of Aerospace Engineering, Delft University of Technology, Rept. LR- 781, Delft, The Netherlands
54. Falcoz A, Henry D, Zolghadri A (2008) A nonlinear fault identification scheme for reusable launch vehicles control surfaces. Int Rev Aerosp Eng 1(5):447–457
55. Fan M, Tits A, Doyle J (1991) Robustness in the presence of mixed parametric uncertainty and unmodeled dynamics. IEEE Trans Autom Control 36(1):25–38
56. Packard A, Doyle J (1993) The complex structured singular value. Automatica 29(1):71–109
57. Newlin M, Smith R (1998) A generalization of the structured singular value and its application to model validation. IEEE Trans Autom Control 43:901–907

Chapter 8
Conclusions and Outlook

In this chapter, we would like to discuss briefly some future challenges and opportunities.

8.1 Fault Detection and Diagnosis

Advanced FDD techniques have probably the strongest potentialities for widespread and real industrial applications in aerospace domain. The following facts allow us to be optimistic for the upcoming years:

- FDD methods and techniques are now well established, and their conceptual and theoretical foundations are well mastered.
- Generally, FDD works in an "open-loop" fashion with respect to the controlled system. So, FDD does not affect the stability and cannot bring the system into a dangerous or diverging configuration. Consequently, it does not act a priori as a brake for the certification which may be required on some vehicles. This "open-loop" aspect is very important for aircraft and space applications. Of course, this depends on how the FDD information is managed by the local or global FDIR system.
- The use of more and more innovative technological solutions in modern spacecraft also introduces new sources of possible failures. The applicability of conventional monitoring techniques is becoming increasingly problematic, and this feature motivates the use of more advanced FDD techniques. While clear-cut failures can be uncovered perfectly by the existing monitoring mechanisms, more subtle and soft drifting-type failures must be detected and isolated by the use of more sophisticated FDD techniques.
- Increasing progress in onboard computational equipment and techniques set ups the scene for the application of more sophisticated and powerful model-based FDD methods.

A. Zolghadri et al., *Fault Diagnosis and Fault-Tolerant Control and Guidance for Aerospace Vehicles: From Theory to Application*, Advances in Industrial Control,
DOI 10.1007/978-1-4471-5313-9_8, © Springer-Verlag London 2014

The academic literature on FDD is now saturated and the effort should be put toward the best suited FDD methods capable of handling the real-world aerospace FDD problems to overcome the Death Valley as discussed in Chap. 2. The applicability gap covers mainly the TRL scales 4 and 5. This aforementioned "application bottleneck" should be overcome through collaborative and more coordinated actions federating academic and industrial actors.

An important issue is the need for clear, systematic, and formalized guidelines for tuning. As discussed in Chap. 2, this aspect is very important and often underestimated in the academic publications. The design method should provide high-level design (tuning) parameters that can be easily used by nonexpert operators. A suitable candidate FDD method for any aerospace application should be able to manage, in a systematic way, stringent operational conditions in terms of trade-offs for FDD specifications, computational burden (memory storage, CPU load), and design complexity. It should also be flexible, or in other words, it should offer a possibility to reuse or to build around it, with adequate design and engineering tools.

Additionally, for aircraft applications and throughout this book, the focus was on the link between FDD and improved structural design. Chapters 3 and 4 highlighted why reliable FDD is important to address this issue that, at first sight, may not seem very obvious. The solutions presented in this book are "full-authority" solutions that try to detect and diagnosis fault events and to make the transition to a degraded mode transparent to the pilot by onboard automatic control/guidance systems reconfiguration, using available onboard control resources. One can look at FDD in a different way by relating it directly to the pilot situation awareness. Situation awareness is a constantly evolving picture of the status of the system state and its environment, which is the most critical aspect for managing a critical situation [1]. Difficulties to understand, in an early stage, the implications of certain system, subsystem, or component failures may prevent or complicate the correct achievement of the flight mission. Today, flight deck represents a highly automated mass of complex systems with which the flight crew has to interact. The increase in automation has shifted the role of the pilot away from hands on flying and more toward the system monitoring. The aircraft internal situation perception relies on existing systems which monitor parameters, detect the error once it occurs, and inform the crew by human machine interface (HMI) concept of sudden alarm. This concept presents especially two limitations for the crew. Firstly, the system status description is not complete: the system health is given by OK/NON OK information which can be not representative of the real status of the system. For example, a system which is not yet in error but is incipiently and slightly drifting is considered as good. Secondly, this approach may create sudden and intensive workload, leading to incorrect decisions during time-critical and complex phases. High and sudden increase of crew workload is due to the error cause knowledge and evaluation of the severity of the abnormal situation and its treatment itself. Note that the crew may make the correct decision based on their picture of the situation, but this situation may be in error and not representative.

Early, anticipated and predictive FDD of possible subsystem problems that are developing during the flight (and which may lead to abnormal aircraft configurations) may help the aircrew react more timely with the situation and take appropriate corrective actions. An appealing direction is predictive monitoring of a flight situation. A flight situation is dynamically impacted by several flight parameters and will be characterized as high-level parameters to be monitored. The task is very challenging because of coupled channels for certain situations and also uncertainty propagation. The challenge is to produce a warning system to inform early the pilot about a slowly drifting flight configuration and to provide him/her with information about root faults and possibly some information about the evolution of the situation and the future trends. Flight configuration management should be performed from a global perspective where the effects of other system states have to be considered. This task is made more difficult if multiple failures occur gradually over a period of time and the consequences of system changes spread throughout the aircraft.

This direction for FDD has not, so far, received much attention and needs further investigations in close collaboration with specialists in cognitive sciences and human machine interface to provide innovating, integrated, and interdisciplinary monitoring solutions.

8.2 Fault-Tolerant Control and Guidance

FTC area has been investigated more recently and took advantage of a number of available results in robust and adaptive control. It is a relatively challenging subject with, at the moment, low support from aerospace industry. Industrial decision makers are generally more skeptical about FTC benefits, although many successful demonstrations are available (see Chap. 2). The reason is mostly related to the fact that any modification to already proven and certified flight control laws is considered to be a critical technological modification, which needs long validation, verification, and certification process. Basically, full-authority automatic reconfiguration solutions can also present several inherent drawbacks in terms of piloting and crew decision making. For instance, for manned space flight and aircraft applications, it is likely to lead to cognitive mismatch, a disparity between the aircrew's mental model of the system and the way the system is really working. If the automatic reconfiguration fails to react to a fault or if the aircrew detect that the system is not behaving correctly, they would not be in a position to take control immediately because the previous actions were masked from them: if the situation needs a fast reaction, the pilot would not have the time to retrace the previous actions of the system and achieve a reasonable degree of situation awareness. As far as human operators will keep the ultimate decision and overall responsibility, further research and investigations should be oriented toward "human in the loop," if the final goal is to improve flight crew go-ahead decisions and actions.

Finally, FTG area is not still sufficiently explored and needs more methodological work. The concept could be very promising for space missions where ground intervention could be too complex, too long or temporarily impossible (i.e., in case of automated operation during a critical phase), and/or too costly. FTG could provide a greater flexibility to account for off-nominal conditions, in situations where FTC is not sufficient (inboard control resources limited after a failure) to recover timely the vehicle. The interaction between FTG and FDD/FTC at system level units also needs more investigations.

8.3 Concluding Remarks

Although the development of advanced model-based FDIR techniques can be considered today as a mature field of research within the academic community, their application to real aerospace world has remained very limited. The design methodology involving feasibility analysis and real-world requirements specification is still relatively immature. A representative problem area remains the lack of an effective process for maturing onboard implementation and certification processes. Aerospace industry needs continuous improvement including insertion of new technologies. This is an excellent driver for further research and development of advanced and model-based FDIR approaches. In order to develop useful and successful solutions for future space and aircraft systems, academic research community should work more closely with industrial practitioners. New ideas can require a long time from conception to exploitation: the time constants for basic research and useful application are certainly quite different. On the other hand, basic research in engineering sciences has always benefited from the influence of applications. A good balance between theory and applications is a win-win situation for everyone and will help overcome emerging technical and scientific obstacles that stand in the way of addressing the new challenges in aerospace domain.

Reference

1. Endsley MR, Garland DJ, Shook RW, Goello J, Bandiero M (2000) Situation awareness in general pilot aviation. Annual report, Year 1, prepared for NASA Ames Research Center. Available on http://www.satechnologies.com/Papers/pdf/GA%20SA.pdf

Index

A. Zolghadri et al., *Fault Diagnosis and Fault-Tolerant Control and Guidance for Aerospace Vehicles: From Theory to Application*, Advances in Industrial Control,
DOI 10.1007/978-1-4471-5313-9, © Springer-Verlag London 2014

U

V

W

Y

MIX
Papier aus verantwortungsvollen Quellen
Paper from responsible sources
FSC® C105338

If you have any concerns about our products,
you can contact us on
ProductSafety@springernature.com

In case Publisher is established outside the EU,
the EU authorized representative is:
Springer Nature Customer Service Center GmbH
Europaplatz 3, 69115 Heidelberg, Germany

Printed by Libri Plureos GmbH
in Hamburg, Germany